The Mammal Orders and their relationships within the Animal Kingdom

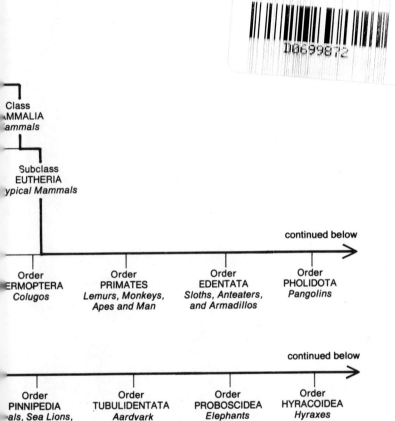

Class
MAMMALIA
Mammals

Subclass
EUTHERIA
Typical Mammals

continued below

| Order ERMOPTERA Colugos | Order PRIMATES Lemurs, Monkeys, Apes and Man | Order EDENTATA Sloths, Anteaters, and Armadillos | Order PHOLIDOTA Pangolins |

continued below

| Order PINNIPEDIA Seals, Sea Lions, Walrus | Order TUBULIDENTATA Aardvark | Order PROBOSCIDEA Elephants | Order HYRACOIDEA Hyraxes |

MAMMALS—Their Latin Names Explained

MAMMALS—
Their Latin Names Explained

A Guide to Animal Classification

A. F. Gotch

BLANDFORD PRESS
Poole   Dorset

First published in the U.K. 1979

Copyright © 1979 Blandford Press Ltd.
Link House, West Street
Poole, Dorset BH15 1LL

British Library Cataloguing in Publication Data
Gotch, Arthur Frederick
    Mammals, their Latin names explained.
    1. Mammals—Nomenclature   2. Zoology—Classification
    I. Title
    599'.001'4   QL708

ISBN 0 7137 0939 1

Phototypeset by Oliver Burridge & Co. Ltd, Crawley, Sussex
Printed in Great Britain by Cox & Wyman, Fakenham

Contents

Acknowledgments

I should like to express my gratitude to the following who have given me help and advice:
L. R. Conisbee; J. Edwards-Hill, British Museum (Natural History); Dr Ernest Neal M.B.E.; Dr J. F. Monk; Dr Hugh Cott; Dr K. A. Joysey; Dr A. Maraspini; Ferelyth and Bill Wills; Michael Gardener; Lesley Lowe; John Maries; Father B. Wrighton; Gordon Bennett; Michael Tweedie; Norman Kearney; Howard Crellin; J. D. Gotch.

Special thanks are due to my friend Ralph Conisbee without whose unfailing help, advice, and enthusiasm, the publication of this book would not have been possible.

I also wish to acknowledge the following publications which were used in research for the book:
Animal Life Encyclopaedia, Purnell & Sons Ltd., London; *Wildlife* (monthly magazine), Wildlife Publications Ltd., London; *A Source-Book of Biological Names and Terms*, E. C. Jaeger, C. C. Thomas, Springfield, Illinois; *Bibliographical Key*, Col. O. E. Wynne, published by Col. O. E. Wynne; *Living Mammals of the World*, I. T. Sanderson, Hamish Hamilton, London; *Living Invertebrates of the World*, R. Buchsbaum and L. J. Milne, Hamish Hamilton, London; *Key to the Names of British Fishes, Mammals, Amphibians and Reptiles*, R. D. Macleod, Pitman, London; *A Classification of Living Animals*, Lord Rothschild, Longmans, London; *The Mammals*, Desmond Morris, Hodder & Stoughton, London; *Principles of Systematic Zoology*, Ernst Mayr, McGraw-Hill, New York; *Index Generum Mammalium*, T. S. Palmer, Government Printing Office, Washington; *Webster's International Dictionary*, G. & C. Merriam Co., U.S.A.; *Grzimek's Animal Life Encyclopaedia*, Van Nostrand-Reinhold, London.

Preface

There is an increasing tendency for books about animals to give their Latin names; this must be welcomed for it enables a species to be identified as *that particular species*, regardless of its English or vernacular name, and it is applicable throughout the world. However, no attempt is made to 'translate' the Latin names into English or to explain the reason for these names.

Some years ago I was studying animals which I had seen in Uganda while on a photographic safari. I was annoyed, because not being a classical scholar, many of the Latin names meant little or nothing to me. After making exhaustive enquiries in bookshops and libraries, including university and zoological libraries, it became apparent that there was no book available that would really solve my problem. Thus I decided, as there was no such book, I had better write one. Actually one or two books of a dictionary type have been published and these might be considered to fulfil the need, but they do not; the names do not refer to any particular species and no attempt is made to explain the *reason* for the names.

There are about 4,300 known species of mammals and I have included over 1,000 of these; most have been selected because they are well known, some because they are of particular interest, and some because they are rare or even on the point of extinction. Some subspecies of particular interest have been included, as they may be ranked as full species by certain authors. Thousands of mammal subspecies, which are sometimes called races, have been named by zoologists. One large mammal, the Virginian Deer *Odocoileus virginianus*, has been divided into nearly forty subspecies, and among the small mammals much greater numbers than these have sometimes been given.

Obviously it would be impossible to include all species of all animals in one book, since such an undertaking would run to many volumes and the research would constitute a lifetime's work for a team of classicists. A Cambridge zoologist carrying out research on beetles worked out that if he spent his entire life studying just one family, the Staphylinidae, he would only be able to devote about twenty minutes to any one species! Each chapter in Part Two of this book (Chapters 6 to 24) refers to one Order, and includes all the Families in that Order, and a diagram shows clearly what animals are included in each Family and how they are classified. Throughout Part 2, the Order, followed by the Suborder where relevant, is shown on the running headline of the left-hand page, and the Family on the right-hand page.

In Part One of the book (Chapters 1 to 5), an explanation is given of the system of classification started by the Swedish naturalist Carl von Linné during the eighteenth century, and which became known as the Binominal System. The basic principles of this system as in use today are explained, avoiding unnecessary detail as far as possible, and throughout the book running headlines at the top of each page give the classification of the animals listed on that page. This makes for quick and easy reference, but the main feature is that with every animal the translation and interpretation of the Latin name is given —and when necessary an explanation of the reason for giving the species that particular name.

In a number of cases reference has been made to T. S. Palmer's standard work *Index Generum Mammalium* 1904 (see page 42); the period from 1904 to 1951 is covered by L. R. Conisbee's British Museum publication *Genera and Subgenera of Recent Mammals*, and from 1951 to 1971 by his four Supplementary Notes in the *Journal of Mammalogy*.

Any reader who is interested in further study of classification and nomenclature should consult the Bibliography given on page 231 at the end of this book.

A. F. Gotch
Summer, 1978

PART ONE
The Animal Kingdom

1 Generic and Specific Names

The student or amateur naturalist who is interested in the study of taxonomy is faced with a formidable array of Latin and Greek words and a system of grouping the animals into phyla, classes, orders and families which have little or no meaning to the layman. In this book my aim is to explain this arrangement and give a meaning to the names by translating them into English; no attempt is made to describe the animal other than the features and associations that explain the reason for the scientific name.

Carl von Linné, the Swedish botanist, born in the year 1707, was responsible for the first real attempts to classify and name living organisms—although Aristotle had done some work on a simple form of classification as long ago as 384 B.C. Linnaeus, the Latin form of the name von Linné, is normally used in classification and his system of nomenclature is now in use throughout the world. The authoritative tenth edition of his *Systema Naturae* was published in 1758, and this is an all-important date, January 1st of that year being the starting point; all Latin names given before that date are considered invalid.

The Linnaean classification is known as the binominal system, i.e. 'two names'; every plant and animal must be given two names. The first is the generic name (from *genus*, which is Latin for birth or origin) and a genus comprises a group of closely related animals (or plants); the second is the specific name, the name of the actual species, which distinguishes it from any other animal in that group, and so from any other animal in the world.

The generic name should be a noun, and written with a capital letter, and the specific name should be an adjective, though sometimes it is a noun; it should be written with a small letter even though it is a proper name of a place or person; and both names should be printed in *italics*. To be completely authentic, it should be followed by the name of the zoologist who first gave the animal that name, and the date of naming; for example the European Hedgehog would be

Erinaceus europaeus Linnaeus, 1758.

The author's name and date should not be in italics; if there has been a change of the generic name, as sometimes happens, the original author's name will be given in parentheses. This information about the author and date is not always necessary in zoological publications and is not given in this book.

PRIORITY

The saying goes 'Priority is the basic principle of zoological nomenclature'; the first scientific name given to an animal after January 1st 1758 stands, even though it is not descriptively accurate. However, by its plenary powers the International Commission on Zoological Nomenclature can moderate the application of this rule to preserve long established names, in order to avoid inconvenience and confusion. A recent ruling has discouraged zoologists from digging out and reviving forgotten names; they must not displace names that have been accepted for fifty years or more.

HOMONYMS

A new name for an animal is considered invalid if it has been used previously for some other form in the *animal* kingdom, as this would result in the same name for two different animals; it is known as a homonym. A difference of one letter is sufficient distinction, for example *Apis*, *Apos* and *Apus* are not homonyms. *Platypus* was given as the Latin name for the Duckbill in 1799, but this was a homonym because the same name had been used for a beetle some years before; it was therefore invalid and a new name had to be given; this was *Ornithorhynchus* (see page 35).

SYNONYMS

In some cases, alternative names are used and recognised. For example, in the primates, an author may use Prosimii and Simiae for the names of the suborders: *pro* (L) before; *simia* (L) an ape, a monkey, i.e. 'early monkeys' or 'primitive monkeys', and 'modern monkeys'; another author may use Lemuroidea and Anthropoidea: *anthrōpos* (Gr) man; *-oides* (New L) from *eidos* (Gr) shape, resemblance, i.e. 'lemur-like' and 'man-like'. These are known as synonyms, and where there are alternative names these will be shown thus: Lemuroidea (or Prosimii), or if there are more than two, Anthropoidea (or Simiae, Pithecoidea): *pithēkos* (Gr) an ape, a monkey. If a generic name has been rejected and considered invalid, but still appears in some publications, it will be shown thus: (*Aotus* formerly *Nyctipithecus*) (see page 84).

TAUTONYMS

It will be seen that in some cases the Latin name of an animal has the generic name repeated for the specific name, for example the Pine Marten is *Martes martes*. This is known as a tautonym, from *tauton* (Gr) the same, and *onoma* (Gr) a name; it has no particular significance and is due to a change of the generic name. For example, a Linnaean specific name later used as a generic name so that the full name is a repeated word; a rule states that the specific name *may not be changed*, even though it results in a tautonym. The ruling has been modified for botanical names and tautonyms are forbidden.

I would like to quote an extract from an article by Mr Michael Tweedie which appeared in the magazine *Animals* (now *Wildlife*):

'The wren was among the birds that Linnaeus himself named, and he called it *Motacilla troglodytes*. Under his genus *Motacilla* he included a number of small birds which ornithologists have now split up into several genera, reserving *Motacilla* for the wagtails. In naming the American wren-like birds in the first decade of the 19th century, the French authority Vieillot chose the Linnaean specific name *troglodytes* (troglodyte or cave-dweller, by reference to the form of the nest) as a generic name. Later it was found that the European wren was so closely allied to these that it must go in Vieillot's genus, so it became *Troglodytes troglodytes*.'

A rule states that the name of the nominate subspecies must repeat the specific name of the species (see Chapter 2). Thus, when subspecies were named, the nominate subspecies of the wren group had to be *Troglodytes troglodytes troglodytes*. The veteran zoologists, who were classical scholars, thoroughly disliked this and there were heated interchanges of correspondence, and they refused to use these 'monstrosities'; but as they passed from the scene the modernists prevailed, the rules were accepted, and so some stability was achieved.

Some authors state that a tautonym indicates the typical or common species in a genus, but this is not correct; it is most unusual for an animal to have a tautonym for its original name. It is true to say that quite often the common species, after a change in its generic name, does then have such a name; for example, the Common Toad *Bufo bufo* and the Common Bat *Pipistrellus pipistrellus*. On the other hand, the names of many common species do not have the tautonymous form; for example, the Common Seal *Phoca vitulina* and the Common Dormouse *Muscardinus avellanarius*.

TYPE SPECIES

The first species to be named in a particular genus is usually, though not always, the type species. Other species in that genus will resemble it more than those in a different genus and they will be distinguished by their specific names. For example, *Martes flavigula* the Yellow-throated Marten, and *Martes americana* the American Marten.

ORIGIN OF NAMES

The Latin name of an animal might originate from the naturalist who first discovered it, but it is more likely that it would originate from a zoologist working in a laboratory, and studying the anatomy of the animal; the specific name is quite often given in honour of the person who discovered it. The Kiwi has been given the generic name *Apteryx*, which is derived from the Greek *a-*, a prefix meaning 'not', or 'there is not', and *pterux* 'a wing'. The kiwi is a flightless bird, the wings being very small, hidden under the body feathers, and useless for flight; the Great Spotted Kiwi *A. haasti* was named in honour of Sir Julius von Haast, the New Zealand explorer.

USE OF LATIN AND GREEK

The advantages of using languages such as Latin and classical Greek are obvious; if Linnaeus had used the Swedish language, then his system would not have been accepted internationally, and in any case other countries would not have understood it. In those days Latin was the international language of European scholars and Linnaeus wrote most of his scientific work in Latin to make it more widely read and understood. Even today, there are classical scholars throughout the world who are familiar with Latin and classical Greek and who understand the meaning of the words. It is true they would find some of the words wrongly construed and incorrectly spelt, and even hybridised, combining a mixture of Latin and Greek to form one word; anathema to the purist. Sometimes they are obscure native words which are known as 'barbarisms'. By international agreement, once these names appear in print and are accepted by the International Commission on Zoological Nomenclature, any mistakes must remain. However, the 1961-63 revised Code permits correction of spelling; at the same time it forbids the use of hyphens,* diacritic marks such as the apostrophe, the diaeresis, and the umlaut —so that *Hyperoödon* becomes simply *Hyperoodon* and *mülleri* becomes *muelleri*.

The classical significance of the names is not of first importance, but some knowledge of Latin and Greek is a help in remembering the names, and will often tell you something about the animal. For instance, *Felis nigripes* must surely mean a cat with black feet; in fact it is the Black-footed Cat of South Africa. The real importance of the names is that they are international fixed labels, identifying a particular species for zoologists throughout the world.

PECULIAR NAMES

Sometimes the names are just invented; the zoologist who named the Kookaburra, or Laughing Jackass as it is sometimes called, was so hard put to it in deciding on a generic name that would distinguish it from other kingfishers, that he used the word *Dacelo*; the generic name of the Common Kingfisher is *Alcedo* which is Latin for a kingfisher; *Dacelo* is simply an anagram of *Alcedo*.

There are a number of other peculiar names apart from anagrams, for instance *Ia io* for a bat, named by Oldfield Thomas; the shortest name ever given to a mammal. Ia was a young woman of classical times and Thomas says: 'Like many women of those times a bat is essentially flighty'; *io* is a Latin exclamation of joy, like hurrah! He obviously had a sense of humour—even if it was a rather chauvinistic one. Another name he invented was *Zyzomys* for an Australian mouse

*The Comma butterfly *Polygonia c-album*, meaning 'many-angled with a white 'c' ', has the honour of being allowed to use the hyphen; probably the only exception! The name refers to a white letter 'C' which is plainly marked on the under surface of the wing.

but he never gave an explanation; perhaps he wanted to make sure that it would appear last in any index!

LOCAL AND PERSONAL NAMES

Many zoologists have used the names of little-known localities, which can make interpretation difficult, and there are references to ancient Greek mythology which may have no special significance. Sometimes the naturalist or collector who discovered the animal is honoured by using his name. The French missionary Père Armand David who discovered a strange Asiatic deer near Peking; this has been given the name *Elaphurus davidianus*. Commemorative names usually have *-i*, *-ii*, or *-iana* added to the complete name to form the genitive if the person is a man, and *-ae*, or *-iae* if a woman. The pheasant *Chrysolophos amherstiae* was named in honour of Lady Amherst; note that in such cases the specific name is written with a small letter in accordance with the rules of nomenclature, although it is a personal name.

In many cases the names are derived from Greek words, but they are 'Latinised', that is to say they are put in a Latin form, and so are referred to as 'Latin names'; it is the popular term but rather disparaged by zoologists who prefer to call them 'scientific names'; purists even suggest that the correct term is 'Linnaean names'.

MISLEADING NAMES

Some animals have been given a name that is misleading when translated; this may have come about because they were named when knowledge of them was incomplete. Here are some examples: *galē* (Gr) a weasel or marten-cat, used for a wallaby, which is a marsupial (see page 46); *meles* (L) a badger, used for the bandicoot, another marsupial (see page 28); *kapros* (Gr) a wild boar, used for the hutiacouga, which is a rodent (see page 139); *galē*, and some other names such as *kuōn* (Gr) a dog, have become conventional final elements in nomenclature for the names of small mammals quite remote from weasel, marten-cat, or dog. The Greek word *mus* (genitive *muos*) a mouse, is used for all kinds of small mammals. Consider, for instance, *Cynomys ludovicianus* the Prairie Dog (see page 118); *Cynomys* means 'dog-mouse', but it is not a mouse and not even in the same family as the rats and mice, and it certainly is not a dog! It will be seen that it is spelt *-mys*, which is the Latin System of transliteration from the Greek; in the Modern System, as used in this book, it is spelt -*mus*. This is the same as the Latin word *mus* (genitive *muris*) which is derived from the Greek and has the same meaning. The two systems of transliteration will be found in the Appendix on page 230.

2   Separating the Animals into Groups

A division or group is known as a taxon (plural taxa) from the Greek word *taxis*, an arrangement, though the term is not often used for groups higher than classes. The divisions are named as shown in the following list, the main divisions in the left-hand column:

KINGDOM
  SUBKINGDOM
PHYLUM
  SUBPHYLUM
  SUPERCLASS
CLASS
  SUBCLASS
  SUPERORDER
ORDER
  SUBORDER
  SUPERFAMILY
FAMILY
  SUBFAMILY
  Genus
    Subgenus
  Species
    Subspecies (or *Race*)

These categories do not all have to be used in the classification of any particular group or species; each division or subdivision is given its own name, as will be seen in Chapter 4, and as shown in the classification key on page 98. A subspecies will have a subspecific name, for example the Helmeted Guinea Fowl *Numida meleagris mitrata*; this is known as trinominal, i.e. three names. It is a subspecies of the Tufted Guinea Fowl *Numida meleagris*. A rule made by the International Commission on Zoological Nomenclature states that the subspecific name of the species on which a group is founded must be the same, i.e. it must *repeat* the specific name. Thus, in this case it becomes *Numida meleagris meleagris*. Therefore it is easily distinguished and is known as the nominate subspecies.

A subspecies is always basically the same as the species, but there is some slight difference; it usually has a different geographical distribution and develops different characteristics. For instance, the crest of the Helmeted Guinea Fowl is larger than that of the Tufted Guinea Fowl and the colour of the body feathers is different. Subgenera are also sometimes recognised, and like subspecies the animal is basically the same as that in the genus; a subgenus is shown thus: *Nycticeius (Scotoecus) falabae* the Falaba House Bat. The subgeneric name is given in parentheses, and does not count as one of the words in the name—so it is not a trinominal name. Some zoologists might consider this bat worthy of the status of a full genus, in which case the name would be *Scotoecus falabae*. Examples of suborders will be found on page 113, and subclasses on page 32 (see also the classification key on pages 98 and 99).

In deciding to which particular taxon an animal belongs, the zoologists must study its anatomy. During the early days of classification not enough attention was paid to this aspect and many animals were placed in the wrong taxon simply because of their outward appearance. These mistakes have gradually been corrected over the years, which accounts for the tautonymic names (see page 13). However, there is always likely to be a difference of opinion among the scientists concerned in this work. One of the most important parts of the anatomy which must be studied is the teeth; during a discussion on the classification of mammals Baron Cuvier, the famous French zoologist, is reputed to have said:

'Show me your teeth and I will tell you what you are.'

Apart from the anatomical structure some notice is taken of the habits of an animal: where does it live, what does it eat? As already mentioned, it is no use looking at an animal and deciding from its appearance to which group it belongs. For instance, take the lizard known in Britain as the 'slow-worm'; the uninitiated on seeing this animal would cry, 'Look, a snake'! Further, many might probably get a stick and beat the poor thing to death, though a more harmless and docile little creature it would be difficult to find (this could well be the reason for the decrease in numbers of the slow-worm in the

British Isles!).

What then makes this creature a lizard and not a snake? For one thing, it has eyelids, and snakes do not have eyelids—though lizards do; and another thing, if caught by the tail the slow-worm, like the lizard, can break it off. This is known as autotomy, from *auto-* (Gr) 'self', and *tomē* (Gr) 'a cutting off'. There is a muscular mechanism in the tail which breaks it off and then actuates to prevent loss of blood; the tail then grows again. This leaves the surprised predator with only the broken-off tail, while the slow-worm makes good its escape. Indeed the Latin name *Anguis fragilis* suggests this, though it also shows that the naturalist who named it thought it was a snake: *anguis* (L) 'a snake', *fragilis* (L) 'brittle'—hence 'the brittle snake'. One can imagine his surprise if he grabbed it by the tail, and was left with it wriggling in his hand; there is no doubt that the dismembered piece does wriggle, as I have seen myself, and this probably adds to the general surprise and confusion and gives more time for the main body to escape.

The main divisions of the Animal Kingdom are the phyla (singular, phylum) from *phulon* (Gr) 'a stock, race or kind'. There is not complete agreement among zoologists as to the number of phyla, as some have separated a group and classified it as a subphylum, and it can take many years before international agreement is reached. As a result a number of different systems have become more or less established and generally recognised, and although they are basically the same there are certain differences which can be confusing; for example one well-known system is based on 22 phyla, and another, as given in this book in Chapter 3, is based on 27 phyla.

In any particular phylum there will be assembled all the animals having a common basic plan. Let us take as an example the phylum Chordata. The distinctive character of this phylum is the notochord; *nōton* (Gr) 'the back' and *khordē* (Gr) 'gut, string', giving rise to *chordata* (New L) meaning 'provided with a back-string'. This is a cord running along the back, made of a special tough elastic tissue, and present in all animals in this phylum. In the humblest members of the group, such as the lancelets (small marine creatures about 5 cm or 2 in long) the notochord is retained throughout life. In the higher forms, the true vertebrates, it is present in the embryo, but is replaced more or less completely by the stronger and yet flexible spinal column of jointed vertebrae. However, as all the animals in the phylum Chordata do not develop a spinal column, and there are other differences, the group is divided into four subphyla, the sub-phylum Vertebrata being one of these; this will be considered in more detail in Chapter 4.

3   The Phyla

As already mentioned, there is not complete agreement about the number of phyla needed to classify the animal kingdom, but the system used in this book comprises 27 phyla, one of which, Echinodermata, has two subphyla, and another, Chordata, four subphyla. Considering that the phylum Chordata includes such diverse animals as the acorn worm, the mouse, the whale and humans, it is no wonder that some subphyla are needed. Even so, the one subphylum Vertebrata contains all the animals best known to us: birds, fishes and reptiles; amphibians, such as the frogs and newts; and mammals, such as dogs, horses, antelopes, lions, whales and ourselves.

I do not advise anyone attempting to learn, like a parrot, the Latin and Greek names that now follow; in the course of study they will become familiar without any special effort. The sequence is not fixed; it begins with what the compiler considers to be the most primitive forms of life and ends with the most advanced. The number of species given for each phylum is approximate; taxonomists do not always agree about the number of species in a particular group, and it can never be finally settled as new species may be discovered at any time.

The word phylum itself is derived from *phulon* (Gr) plural *phula*, a race or kind. The phyla are as listed below.

PROTOZOA   30,000 species. Amoeba, mycetozoa, etc.
prōtos (Gr) first; *zōon* (Gr) an animal, a living thing.
amoibē (Gr) a change, an alteration. The amoeba, a tiny single-celled animal, continually changes its shape.
mukēs (Gr), genitive *mukētos*, a fungus; any knobbed body shaped like a fungus.

PORIFERA (or PARAZOA, SPONGIDA) 4,500 species. The sponges.
poros (Gr) a way through, a passage; *gero* (L) I bear, I carry.

COELENTERATA (or CNIDARIA) 9,000 species. Jellyfish, corals, etc.
koilos (Gr) hollow; *enteron* (Gr) bowel, intestine.

CTENOPHORA 80 species. Comb jelly, sea gooseberries.
kteis (Gr), genitive *ktenos*, a comb; *phora* (Gr) carrying, bearing.

MESOZOA 7 species. Minute worms, parasites in the kidneys of squids and octopuses.
mesos (Gr) middle; *zōon* (Gr) an animal, a living thing.

PLATYHELMINTHES 9,000 species. Flatworms, liver flukes, bilharzia, etc.
platus (Gr) flat; *helmins* (Gr), genitive *helminthos*, a worm.

NEMERTEA 570 species. Ribbon worms, bootlace worms.
nēma (Gr) genitive *nēmatos*, a thread.

NEMATODA (or NEMATA) 10,500 species. Roundworms, vinegar eelworms, etc.
nēma (Gr), genitive *nēmatos*, a thread; *-oda* (New L) from *eidos* (Gr) form, like.

ROTIFERA (or ROTATORIA) 1,200 species. Wheel animalcules, the smallest many-celled animals about 1 mm (less than $\frac{1}{16}$ in) long, mostly living in fresh water.
rota (L) a wheel; *fero* (L) I bear, I carry. They do not, of course, carry a wheel, but the first ones discovered and examined under a microscope showed tiny hairs round the mouth; these wave in a circular motion that gives the impression of a turning wheel.

GASTROTRICHA 100 species. Gastrotrichs, tiny transparent creatures less than 0·5 mm ($\frac{1}{50}$ in) long, mostly living in fresh water.
gaster (Gr) the belly, stomach; *thrix* (Gr), genitive *trichos*, hair; they have hairs on their underparts.

KINORHYNCHA (or ECHINODERA, ECHINODERIDA) 30 species. Kinorhynchs, marine animals about 1 mm (less than $\frac{1}{16}$ in) long.
kineō (Gr) I move; *rhunkhos* (Gr) a snout, a beak; they pull themselves along by a kind of snout.

PRIAPULIDA 6 species. Priapulids, wormlike marine animals 7·6 cm (about 3 in) long, living on muddy bottoms and sometimes at a great depth. Priapos was the god of gardens and vineyards and Priapus, in Roman mythology, meant a representation or symbol of the male generative organ, or phallus. The name priapulida probably refers to the shape of the animal, which is not unlike the human penis; -*ida* (New L) from *idea* (Gr) species, sort.

NEMATOMORPHA (or GORDIACEA) 80 species. Horsehair worms, not usually marine, and very variable in length from 10 cm (4 in) upwards, and from 0·3 to 3 mm ($\frac{1}{80}$–$\frac{1}{8}$ in) in diameter.
nēma (Gr) genitive nēmatos, thread; *morphē* (Gr) form shape. Sometimes found in horse drinking troughs, hence the English name.

ACANTHOCEPHALA 400 species. Spiny-headed worms; parasites in the intestines of vertebrates.
akantha (Gr) a thorn, prickle; *kephalē* (Gr) the head.

ENTOPROCTA (or ENDOPROCTA, CALYSSOZOA, KAMPTOZOA, POLYZOA ENDOPROCTA, POLYZOA ENTOPROCTA) 60 species. Entoprocts; a flower-like body on a stalk less than 6 mm ($\frac{1}{4}$ in) high. The anus opens within a circlet of tentacles, hence the name: *entos* (Gr) within; *prōktos* (Gr) anus, hinder parts.

CHAETOGNATHA 30 species. Arrow worms; common in sea water near the shore or in the depths, from 2 to 10 cm ($\frac{3}{4}$–4 in) in length.
khaitē (Gr) hair, mane; *gnathos* (Gr) jaw; it has short bristles surrounding the mouth.

POGONOPHORA (or BRACHIATA) 22 species. Beard worms; a deep sea animal about 3 mm ($\frac{1}{10}$ in) in diameter and up to 33 cm (13 in) long. It has no digestive system and how it obtains nourishment is something of a mystery.
pōgōn (Gr) a beard; *phora* (Gr) carrying.

PHORONIDA (or PHORONIDEA) 15 species. Phoronids; small marine animals that range from 2 to 30 cm (1 –12 in) in length. They build themselves a tube in which to live, and catch their food by means of tentacles projecting from the end of the tube. The reason for their name is obscure, but is thought to originate from Phoronis, surname of Io, the daughter of Inachus; there is a strange legend which involves her wandering all over the earth.

BRYOZOA (or POLYZOA, POLYZOA ECTOPROCTA, ECTOPROCTA) 6,000 species. Moss animals; tiny animals, about 0·4 mm ($\frac{1}{64}$ in) long, which live in sea water and fresh water. You might find them attached to the bottom of your boat when you pull it ashore, and though they look rather like a mossy plant they are probably these tiny animals, the bryozoans.
bruo (Gr) I swell, I sprout, giving rise to bryon, lichen or tree moss; *zōon* (Gr) an animal.

BRACHIOPODA 260 species. Lamp shells; the brachiopod shell is shaped somewhat like the oil lamps used by the Greeks and Romans in ancient times.
brakhus (Gr) short; *pous* (Gr) genitive podos, a foot; they do not have feet, but this refers to a short stalk by which they attach themselves to some support.

SIPUNCULA (or SIPUNCULOIDEA) 250 species. Peanut worms; marine animals about 20 to 40 cm (8-18 in) long and 12 mm ($\frac{1}{2}$ in) in diameter. They can change their shape, and sometimes take the shape of a peanut. A peculiar feature is a pump-like action with one part of the body sliding up and down inside the posterior cylindrical part.
siphōn (Gr) a sucker, as of a pump; *-culus* (L) suffix meaning small: 'a little pump'.

ECHIURA (or ECHIUROIDEA, ECHIURIDA) 60 species. Curious sausage-shaped marine animals, up to about 30 cm (12 in) long, and having a proboscis which in some species may be 1 m (3 ft) long. This is used for gathering food: *ekhis* (Gr) an adder, a serpent: *oura* (Gr) the tail; the 'serpent' part is more a nose than a tail.

MOLLUSCA 50,000 species. Oysters, octopuses, slugs, snails, squids, etc.
mollusca (L), neuter plural of *molluscus*, soft.

ANNELIDA (or ANNULATA) 6,000 species. Leeches, earthworms, ragworms, etc.
anellus (L) a little ring; *-ida* (New L) from *idea* (Gr) species, sort; the rings that mark the body of an earthworm or a ragworm give this phylum its name.

ARTHROPODA 815,000 species, including the insects which number about 750,000 species, possibly many more. Crustaceans, spiders, insects, etc.
arthron (Gr) a joint; *pous* (Gr) genitive podos, a foot; in this case it is taken to mean leg, as the arthropods have jointed legs. Some also have a hard external skeleton; *crusta* (L) a shell, crust.

ECHINODERMATA 5,500 species.
ekhinos (Gr) a hedgehog; *derma* (Gr) the skin; a reference to some that have 'spiky' skins.

Subphylum Pelmatozoa Sea lilies, feather stars.
pelma (Gr) genitive *pelmatos*, the sole of the foot, a stalk; *zōon* (Gr) an animal. The sea lily has a flower-like body supported on a stalk, the feather star begins life on a stalk but later breaks away and is free to swim about.

Subphylum Eleutherozoa Sea urchins, sea cucumbers, starfishes etc.
eleutheros (Gr) free, not bound as on a stalk; *zōon* (Gr) an animal: the sea urchin's body is equipped with movable spikes.

CHORDATA about 44,750 species.
nōtos (Gr) the back; *khordē* (Gr) gut, string; *-ata* (New L) suffix used for divisions of zoological names; the notochord, or 'back-string', is a rod-like structure made of tough elastic tissue which is present in all early embryos in the phylum Chordata.

Subphylum Hemichordata (or Stomochordata, Branchiotremata) 90 species. The acorn worms and their kin.
hēmi- (Gr) prefix meaning half; suggesting halfway between primitive chordates and the next stage. The notochord is found only in the proboscis.

Subphylum Urochordata (or Tunicata) 1,600 species. Sea squirts, salps, etc.
oura (Gr) the tail; *khordē* (Gr) gut, string. The notochord extends into the tail.

Subphylum Cephalochordata (or Acrania, Leptocardii) 13 species. Lancelets; small marine animals about 5 cm (2 in) long and pointed at both ends.
lancea (L) a small spear; *kephalē* (Gr) the head: *khordē* (Gr) gut, string. The notochord extends into the head.

Subphylum Vertebrata About 43,000 species. The vertebrates; fishes, amphibians, reptiles, birds, and mammals, including humans (see page 28).

vertebra (L) a joint, specially a joint of the back, derived from *verto* (L) I turn. With the exception of certain fishes, such as the cartilage fishes, during development from the embryo the notochord is replaced by the bony spinal column.

4   The Vertebrates*

In the subphylum Vertebrata there is a great variety of animals, and so they are divided into groups, or taxa, called classes. There are certain recognised sequences for compiling lists of animals, but they are arbitrary, and different authors adopt different plans. It is not possible to make a linear series that is scientifically correct from the point of view of evolution, because mammals, or any other forms of life, have not descended one from another in a long line. The classes are as listed below.

MARSIPOBRANCHII (or AGNATHA)   Lampreys and hagfishes.

SELACHII (or CHONDROPTERYGII, CHONDRICHTHYES, ELASMOBRANCHII)   Sharks, dogfishes, rays, i.e. the cartilage fishes.

BRADYODONTI   Rabbit fishes.

PISCES (or OSTEICHTHYES)   Bony fishes.

AMPHIBIA   Frogs, toads, newts, and their kin.

REPTILIA   Tortoises, lizards, snakes, crocodiles.

AVES   Birds.

MAMMALIA   Dogs, cats, horses, whales, man, and their kin.

The mammals are divided into 19 taxa called orders. They show the most amazing variation in form, and have adapted themselves to live in the air, in the water, on the ground, and under the ground.

*Including the cartilage fishes.

The orders are:

Monotremata	Duck-billed platypus, and echidna or spiny ant-eater
Marsupialia	Kangaroos, opossums, and their kin
Insectivora	Hedgehogs, shrews, moles, and their kin
Chiroptera	Bats
Dermoptera	Colugos or flying lemurs
Primates	Tree shrews, lemurs, monkeys, apes and man
Edentata	Sloths, armadillos, anteaters
Pholidota	Pangolins
Lagomorpha	Rabbits, hares, pikas
Rodentia	Mice, rats, squirrels, porcupines, and their kin
Cetacea	Whales, dolphins, porpoises
Carnivora	Lions, dogs, mongooses, weasels, and their kin
Pinnipedia	Seals, sea-lions, walrus
Tubulidentata	Aardvark
Proboscidea	Elephants
Hyracoidea	Hyraxes
Sirenia	Manatees, dugong
Perissodactyla	Horses, tapirs, rhinoceroses
Artiodactyla	Deer, camels, giraffes, hippopotamuses, and their kin

The orders are divided into families and the family taxon is easy to distinguish because the scientific name ends with -idae and a sub-family with -inae. It is formed by adding the suffix -idae, or -inae, to the stem of the name of the type-genus. For example, the family name of the squirrels is formed from the squirrel genus *Sciurus*, and so becomes Sciuridae. The type-genus is that genus whose structure and characteristics are most representative of the larger group as a whole, although in some cases it may have been selected because it is the largest, best-known, or earliest-described genus.

To bring us up to date so far we will list the divisions, starting with the phylum, which apply to the Perissodactyla:

Phylum   CHORDATA
Subphylum VERTEBRATA
Class   MAMMALIA

Order   PERISSODACTYLA

Families   EQUIDAE   Horses
TAPIRIDAE   Tapirs
RHINOCEROTIDAE   Rhinoceroses

Now we come to the genera and species, which will give each animal its own name.

Family EQUIDAE
Genus and species *Equus przewalskii* Przewalski's horse
E. caballus   Horse
E. hemionus   Wild ass
E. asinus   North African wild ass (donkey)
E. zebra   Mountain zebra
E. burchelli   Burchell's or common zebra
E. grevyi   Grèvy's zebra

Family TAPIRIDAE
Genus and species *Tapirus indicus* Malayan tapir
T. terrestris   Brazilian tapir
T. pinchaque   Mountain tapir
T. bairdi   Baird's tapir

Family RHINOCEROTIDAE

Genus and species *Rhinoceros unicornis*   Indian rhinoceros
R. sondaicus   Javan rhinoceros
Didermocerus sumatrensis Sumatran rhinoceros
Diceros simus   White rhinoceros
D. bicornis   Black rhinoceros

It is standard practice, where the genus is the same as the one mentioned immediately before, simply to put the initial capital letter. This practice has been followed in the lists above. The English interpretation of all the foregoing Latin names will be given in the Chapter dealing with each particular genus.

Phylum   Chordata
Subphylum   Vertebrata

AN EXAMPLE

Now let us look at an example of a small mammal, the House Mouse, with translations into English of each stage of its classification.

khorde (Gr) gut, string, giving rise to chordata (New L) having a notochord, or backstring: *notos* (Gr) the back. The notochord is made of a special tough elastic tissue, and is present in the embryo of all animals in this Phylum. In the Subphylum Vertebrata it develops into the spinal column, as in this case.

vertebra (L) a joint, specially a joint of the back, derived from *verto* (L) I turn.

Class   Mammalia

mamma (L) a breast. All animals which feed their young from the breast are in this class, so it includes the whales, seals and other sea-living mammals.

rodo (L) I gnaw, eat away.

Order   RODENTIA

Family   Muridae   *mus* (L) genitive muris, a mouse or a rat: *-idae* (L) suffix added to generic names to form family names.

Genus   *Mus*

Species   *musculus*   *-culus* (L) suffix added to form diminutive. So the House Mouse is known as *Mus musculus*—'Mouse, little mouse'.

5 The Mammals

This group of animals—the class Mammalia—includes tiny creatures like the harvest mouse, weighing about 7 g (only $\frac{1}{4}$ oz), and the whales which may weigh over 100 tonnes. Between these two extremes are all the animals most familiar to us—such as cats, dogs, horses, lions, elephants and ourselves. They are the only animals that have true hairs and, by means of the mammary glands produce milk to feed their young; hence the name 'mammals', the word being derived from the Latin *mamma*, a breast.

For purposes of classification the class Mammalia is divided into three subclasses: the Prototheria or 'first animals', the Metatheria or 'later animals', and the Eutheria, 'typical' or 'well-made animals'. The echidnas and the duck-billed platypus are the only ones in the subclass Prototheria, the kangaroos and other pouched animals form the subclass Metatheria, and all the remainder form the subclass Eutheria.

The first group, the Monotremes, are primitive animals that lay eggs. The second group, the Marsupials, are a more advanced form of life in which the young are born after a very short gestation period, but are not fully developed and have to start life in the pouch. All the animals in the third group give birth to young in an advanced state of development, though in most cases the young are quite helpless and dependent on the care of the mother. For instance, a lioness may continue to teach her cubs to hunt for about a year.

The placenta, which unites the foetus to the mother's womb and supplies it with nourishment and oxygen, is not found in the egg-laying mammals, and only in a rudimentary form in the marsupials. All the remaining mammals have the placenta, and it is not found in any other animals (except for a rather similar organ in one or two fishes and lizards). It has almost certainly evolved from the reptiles

PART TWO
The Mammal Orders

6   Duck-Billed Platypus and Echidnas
MONOTREMATA

Subclass   PROTOTHERIA
prōtos (Gr) first; *thēr* (Gr) a wild animal.

Order   MONOTREMATA
monas (Gr) single; *trēma* (Gr) a hole.

Family   ORNITHORHYNCHIDAE   one species
ornis (Gr) genitive *ornithos*, a bird; *rhunkhos* (Gr) a beak, a bill.

Duck-billed Platypus   *Ornithorhynchus anatinus*
anas (L), genitive *anatis*, a duck   *-inus* (L) suffix meaning like,
belonging to. Platypus is from *platus* (Gr) flat, broad and *pous* (Gr) a
foot. This is the only known species; it is a most peculiar animal, and
has some similarity to birds as well as mammals, and even to reptiles.
It is one of the very few mammals that is truly venomous; the male
has sharp spurs on the hind legs that are connected by a duct to
poison glands. They have only one aperture for the elimination of
liquid and solid waste, for copulation, and for the birth of the young;
hence the name 'monotreme'. They lay soft-shelled eggs in a nest like
the reptiles. Found only in Australia and Tasmania.

Family   TACHYGLOSSIDAE   five species
takhus (Gr) fast, swift   *glossa* (Gr) the tongue.

Echidna or Australian Spiny Anteater   *Tachyglossus aculeatus*
aculeus (L) a sting, a point; *aculeatus* thus means provided with prickles

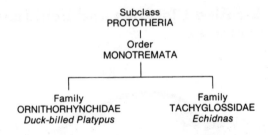

Subclass
PROTOTHERIA
|
Order
MONOTREMATA

Family
ORNITHORHYNCHIDAE
Duck-billed Platypus

Family
TACHYGLOSSIDAE
Echidnas

or stings. This probably refers to the body, not to the tongue, or to the spurs on the hind legs. These spurs are equipped with a poison duct, hence *echidna* (Gr) a viper, an adder (see Platypus above). With the echidna, unlike the platypus, the eggs remain in a pouch which develops after mating; later the mammary glands discharge milk into this pouch.

Tasmanian Spiny Anteater   *T. setosus*
saeta (= seta) (L) a bristle   *-osus* (L) suffix meaning full of.

New Guinea or Three-toed Spiny Anteater   *Zaglossus bruijni*
za (Gr) prefix with intensive sense, meaning very, much   *glossa* (Gr) the tongue; 'a long tongue'. A. A. Bruijn (fl. 1875–1885) was an officer in the Dutch navy; he collected specimens in the Malay Archipelago, especially Celebes and New Guinea.

7 Marsupials or Pouched Animals
MARSUPIALIA

In this group of mammals the young are born very small, and as little more than partly developed embryos; but they quickly find their way to the mother's pouch. Although collectively known as 'pouched mammals', in some cases the pouch is only a fold of skin and in others is completely absent. The young then attach themselves to a teat and remain attached for several weeks.

Subclass METATHERIA

meta (Gr) among, between; of place, after; of time, after, later   *thēr* (Gr) a wild animal (see page 32).

Order MARSUPIALIA

marsupium (L) a purse, a pouch.

Family DIDELPHIDAE   about 65 species
di- (Gr) prefix meaning two, double   *delphus* (Gr) the womb; an allusion to the pouch as a 'secondary womb' in which the young develop after birth.

Woolly Opossum   *Caluromys lanatus*
kalos (Gr) beautiful, fair   *oura* (Gr) the tail   *mus* (Gr) a mouse   *lana* (L) wool   *-atus* (L) suffix meaning provided with, i.e. 'woolly'. Inhabiting tropical South America.

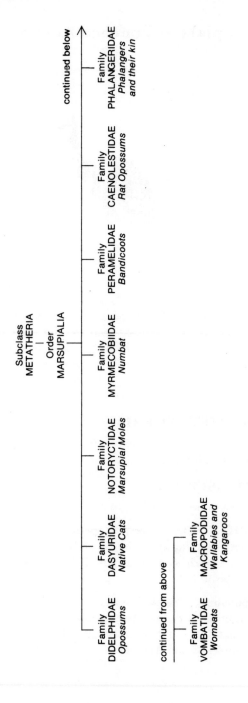

Common or Virginia Opossum   *Didelphis marsupialis*
-*alis* (L) suffix meaning pertaining to, like. Inhabiting North and
South America.

Murine or Mouse Opossum   *Marmosa murina*
marmose (Gr) a name of obscure origin, given by the Comte de Buffon
in the 18th century   *mus* (L), genitive *muris*, a mouse   *ina* (L) suffix
meaning like, belonging to. Inhabiting Brazil. This is one of a very
large genus of about 40 species.

Water Opossum or Yapok   *Chironectes minimus*
kheir (Gr) the hand   *nēo* (Gr) I swim   *nēktēs*, a swimmer; 'a hand for
swimming'. They have webbed feet and are good swimmers   *minimus*
(L) smallest; it was first described as an otter, and in comparison it
was very small. The name Yapok is derived from the South American
river Oyapok, between Brazil and Guyana. Inhabiting Guatemala
and southern Brazil.

Family   DASYURIDAE   about 45 species.
dasus (Gr) hairy   *oura* (Gr) the tail; they have a bushy tail.

Broad-footed Marsupial Mouse   *Antechinus stuarti*
anti- (Gr) prefix meaning against, opposed to, but can also mean
equal to, like, resembling the word that follows   *echinos* (Gr) a
hedgehog; the head is similar to a hedgehog but it does not have
prickly spines. J. McDouall Stuart (1815-1866) was a Scottish-
Australian explorer who made several expeditions to the interior;
Mount Stuart is named after him. The small marsupials are indeed
mouse-sized, but they are not mice. Widespread in Australia.

Flat-skulled Marsupial Mouse   *Planigale ingrami*
planus (L) level, flat   *galea* (L) a helmet; the skull is so flattened that
it can creep through 6 mm ($\frac{1}{4}$ in) crevices. It is the smallest marsupial
and is named after Sir William Ingram, Bart (1847-1924), an English
newspaper proprietor who shared in the financing of a collecting
expedition in Australia. Named by Oldfield Thomas in 1906, it
inhabits Western Australia. Oldfield Thomas (1858-1929) was at
one time curator of mammals at the Natural History Museum,
London. His classic paper of 1911 identified the Linnaean mammals
of 1758 and is a landmark in mammalian nomenclature. He probably
wrote more papers on and named more genera and species of mammals than any other zoologist, during the years 1880 to 1929. He

came to a tragic end, as he took his own life through grief at his wife's death.

Crest-tailed Marsupial Mouse   *Sminthopsis crassicaudata*
sminthos (Gr) an old Cretan word for a field mouse   *opsis* (Gr) aspect, appearance   *crassus* (L) thick   *cauda* (L) the tail of an animal   *-atus* (L) suffix meaning provided with. Inhabiting central and southern parts of Australia.

Pouched Jerboa   *Antechinomys spenceri*
Antechinus, see above   *mus* (Gr) a mouse. Named in honour of Professor Sir W. B. Spencer (1860–1929), a Director of the Natural History Museum in Melbourne and an explorer. He discovered *Notoryctes* (below). Inhabiting the desert areas of Australia.

Eastern Dasyure or Quoll   *Dasyurus quoll*
dasus (Gr) hairy   *oura* (Gr) the tail; it has a bushy tail. Quoll is an Aborigine word for this animal; it lives in Australia and Tasmania.

Western Native Cat   *Dasyurinus geoffroii*
-inus (L) suffix meaning like, belonging to. Named after Etienne Geoffroy Saint-Hilaire (1772–1844) the well known French zoologist. His son Isidore was also a zoologist and wrote a book about the life and work of his father. This dasyure inhabits the western part of Australia.

Little Northern Dasyure or Native Cat   *Satanellus hallucatus*
Satan (Gr) satan, from a Hebrew word meaning 'the enemy'   *-ellus* (L) diminutive suffix; 'little devil'; probably referring to its destructive nature   *hallex* (L), genitive *hallicis*, the thumb or big toe   *-atus* (L) suffix meaning provided with. Unlike other dasyures this one has the big toe. Inhabiting northern Australia.

Tiger Cat or Spotted Native Cat   *Dasyurops maculatus*
Dasyurus, see above; *ops* from *opsis* (Gr) aspect, appearance; 'like a dasyure'   *macula* (L) a spot, a mark   *-atus* (L) suffix meaning provided with, i.e. spotted, speckled; it has white spots on the body. Widespread in Australia.

Tasmanian Devil   *Sarcophilus harrisii*
sarx (Gr), genitive *sarkos*, flesh   *philos* (Gr) loved, pleasing; 'a meat lover'. Named after Deputy Surveyor Robert Harris; his notes on this animal were published in London in 1808. It is probably now confined to Tasmania.

Thylacine or Pouched Wolf *Thylacinus cynocephalus*
thulakos (Gr) a bag, pouch *cinus* (=*cynus*) from *kuōn* (Gr), genitive
kunos, a dog *kephalē* (Gr) the head; it has a head like a dog or wolf.
Inhabiting the mountains of Tasmania, and now very rare, possibly
extinct.

Family NOTORYCTIDAE 2 species

Marsupial Mole *Notoryctes typhlops*
notos (Gr) the south *oruktēs* (Gr) one who digs; 'a southern digger'
tuphlos (Gr) blind *ops* from *opsis* (Gr) appearance, 'apparently
blind'; the eyes are almost useless and hidden under the fur. It lives
in Australia, hence 'the south'.

Family MYRMECOBIIDAE 1 species

Numbat or Marsupial Anteater *Myrmecobius fasciatus*
murmēx (Gr), genitive *murmēkos*, an ant *bios* (Gr) living, also manner
of living; 'one that lives on ants' *fascia* (L) a band, a girdle *-atus*
(L) suffix meaning provided with; it has dark bands round the body.
Numbat is a name used by the Aborigines of Australia, where this
animal lives.

Family PERAMELIDAE about 22 species

Long-nosed Bandicoot *Perameles nasuta*
pera (L) a bag, a pouch *meles* (L) a badger *nasutus* (L) having a
large nose; it inhabits the eastern part of Australia. Pandikokku,
which became 'bandicoot', is the Telugu word for pig-rat, and used
in southern India and Ceylon for a large rat. The bandicoot-rat of
India almost certainly was familiar to some of the early settlers who
arrived in Australia. Thus there was the transfer of the name to
similar animals that they encountered, even though they were
marsupials and not placentals.

Rabbit Bandicoot *Thylacomys lagotis*
thulakos (Gr) a bag, a pouch *mus* (Gr) a mouse *lagos* (Gr) a hare
ous (Gr), genitive *ōtos*, the ear; 'a pouched mouse with hare's ears'.
Inhabiting Australia.

Brindled Short-nosed Bandicoot *Isoodon marcourus*
isos (Gr) equal, similar *odōn* (Gr) a tooth; the teeth are similar in

size and shape   *makros* (Gr) long   *oura* (Gr) the tail; it has a short
nose and a long tail. Inhabiting the eastern part of Australia.

Pig-footed Bandicoot   *Chaeropus ecaudatus*
khoiros (Gr) a young pig   *pous* (Gr) a foot; alluding to the striking
resemblance of the forefeet to those of a pig   *ex* (L) from out of,
without   *cauda* (L) the tail of an animal   *-atus* (L) suffix meaning
provided with; 'not provided with a tail'. The name is not correct as
the bandicoot *has* a tail, but the first one discovered and named had
no tail, a loss to which this animal is apparently prone. In 1857
J. L. G. Krefft, the Australian zoologist, found that he had great
difficulty in obtaining specimens. He showed a drawing of the original
one with no tail to the natives, who failed to recognise it as their
tailed 'landwang' and brought to him examples of the common
bandicoot minus the tail which they had removed, hoping to satisfy
his requirements. The Pig-footed Bandicoot inhabits southern
Australia and Tasmania, but is now extremely rare, and possibly
extinct.

Family   CAENOLESTIDAE   7 species

Rat Opossum   *Caenolestes fuliginosus*
There tends to be confusion about the two names 'opossum' and
'possum'; there is no significance in the different names, though
'possum' was probably first used in America for the opossums of the
family Didelphidae. Captain James Cook was said to have first used
this abbreviated form and it is sometimes used in Australia for those
in the family Phalangeridae. The expression 'playing possum' alludes
to the animal's habit of lying on its back and pretending to be dead
when in danger from predators.
kainos (Gr) new, recent   *lēstēs* (Gr) a robber   T. S. Palmer, in his
Index Generum Mammalium, quotes—'The affix *lestes* is connected in
mammalogy with small and ancient fossil marsupials, . . . so that the
above may be considered to represent an existing animal with ancient
fossil relatives. (Thomas)'   *fuligo* (L) genitive *fuliginis*, soot   *-osus*
(L) suffix meaning full of, prone to; the coat is dark grey. Inhabiting
Ecuador, South America. Dr Theodore Sherman Palmer (1868–1958)
was at one time Secretary of the American Ornithologists Union;
his *Index Generum Mammalium*, published in 1904, took twenty years to
complete. The work was begun in 1884 by Dr C. Hart Merriam (1855–
1942), Chief of the U.S. Biological Survey, and so pressure of other

work forced him to hand over the task to Palmer. Assisted by a large number of researchers he finally completed the work in 1904, its year of publication.

Chilean or Fat-tailed Rat Opossum   *Rhyncholestes raphanurus*
rhunkhos (Gr) beak, snout   *lēstēs* (Gr) a robber; see above   *raphanos* (Gr) a radish   *oura* (Gr) the tail; the tail is thickened at the base giving it somewhat the appearance of a radish. Very little is known about this rare opossum. It inhabits Chiloe Island, Chile, and the neighbouring mainland.

Family   PHALANGERIDAE   about 46 species
phalanx (Gr), genitive *phalangos*, a line or order of battle; also, in the biological sense, the bone of a finger. Commonly known as phalangers, these marsupials have an unusual dexterity with their fingers, and the big toe is opposed to the other toes as our thumb is opposed to the fingers. However the name is not derived from this; it is an allusion to the peculiarity of the hind foot, in which the second and third digits are webbed together. It has been stated that the name is derived from the Greek *phalanx*, meaning a web. *Phalanx* does not mean a web, but there is a Greek word *phalangion*, a venomous spider, and after the classical period this may also have been used to mean a web.

New Guinea Spotted Cuscus   *Phalanger maculatus*
Phalanger, see above   *macula* (L) a spot   *-atus* (L) suffix, meaning provided with. Cuscus is a native name from the Moluccas islands in Indonesia.

Celebes Cuscus   *P. celebensis*
-ensis (L) suffix meaning belonging to, usually local names; in this case from Celebes.

Brush-tail Opossum   *Trichosurus vulpecula*
thrix (Gr), genitive *trikhos*, the hair, of man or beast   *oura* (Gr) the tail   *vulpes* (L) a fox   *-culus* (L) diminutive suffix; 'a little fox'. Inhabiting Australia, Tasmania, and New Zealand.

Pygmy Flying Possum or Pygmy Glider   *Acrobates pygmaeus*
akros (Gr) at the top   *bates* (Gr) from *baino*, I walk, step; 'an acrobat' *bugmaios* (Gr) dwarfish. Flying possums have a web of skin stretching from the forepaws to the hind paws which enables them to glide quite

long distances: see notes about possums and opossums on page 42. Inhabiting Australia and New Guinea.

Pentailed Phalanger   *Distoechurus pennatus*
dis (Gr) twice, double   *stoikhos* (Gr) a row, a line   *oura* (Gr) the tail; 'a two-lined tail'   *penna* (L) a feather or wing   *pennatus* (L) feathered; the tail has two lines of long stiff hairs, on both sides, and only very short fur above and below, which gives it the appearance of a feather. Inhabiting New Guinea.

Pygmy Possum or Dormouse Phalanger   *Cercartetus nanus*
kerkos (Gr) the tail   *artaō* (Gr) I fasten to, I hang one thing upon another   *-etus* (L) a suffix usually used to designate a place; it has a prehensile tail, it can hang itself up on its tail   *nanus* (L) a dwarf. Inhabiting Australia, Tasmania, and New Guinea.

Dormouse Possum   *Burramys parva*
Burra is a town in south-eastern Australia where fossil remains of this animal were found in 1890; it was not until 1966 that one was caught alive. *mus* (Gr) a mouse   *parvus* (L) minor.

Leadbeater's Possum   *Gymnobelideus leadbeateri*
gumnos (Gr) naked   *belos* (Gr) anything thrown, a dart   *-ideus* (New L) suffix denoting similarity; 'naked dart' because it lacks the flying membranes of the Flying Possum (see page 43). J. Leadbeater (fl. 1852–1875) was an assistant at the National Museum in Melbourne; he discovered this possum in a forest near Melbourne, Australia.

Sugar Squirrel or Sugar Glider   *Petaurus australis*
petaurum (L) a springboard used by acrobats   *australis* (L) southern; it does not necessarily mean Australia. This Sugar Glider inhabits Tasmania and eastern parts of Australia.

Striped Possum   *Dactylopsila trivirgata*
daktulos (Gr) a finger; can mean a toe   *psilos* (Gr) naked; the toes of this possum are naked   *tres*, or *tria* (L) three   *virgatus* (L) striped. Inhabiting Australia.

Scaly-tailed Possum   *Wyulda squamicaudata*
Wyulda is the West Australian Aborigine name for the brush-tailed possum   *squama* (L) a scale of an animal   *cauda* (L) the tail of an animal. Inhabiting Australia.

Koala Bear   *Phascolarctos cinereus*
phaskōlos (Gr) a leather bag, a pouch   *arktos* (Gr) a bear; 'a bea

with a pouch' *cinis* (L), genitive *cineris*, ashes; cinereus, ash-coloured. These harmless little animals have been sadly persecuted for their fur, but a few colonies are still living in eastern Australia. Koala is an Aboriginal name and is said to mean 'no drink animal'.

Common Ringtailed Opossum *Pseudocheirus laniginosus*
pseudēs (Gr) false, deceptive *kheir* (Gr) the hand; an allusion to the hand-like forefeet, the two inner toes being opposable to the other three *lana* (L) wool and hence *laniger* (L) wool-bearing *-osus* (L) suffix meaning full of; 'woolly'. Inhabiting Australia.

Rock-haunting Ringtailed Opossum *Petropseudes dahlii*
petra (Gr) a rock *pseudēs* (Gr) false, deceptive; when naming this opossum in 1923 Thomas used *petra* because of its rocky habitat, and 'tacked' this word on to *pseudes* to show its relationship with *Pseudocheirus* (above). Named in honour of Professor K. Dahl (1871–1951), a Norwegian zoologist who was in Australia from 1894 to 1896. This opossum is widespread in Australia.

Great Glider *Schoinobates volans*
skhoinobatēs (Gr) a rope-dancer *volo* (L) I fly; *volans*, flying; a large glider that may measure more than 1 m (3 ft) in length. Inhabiting the forests of eastern Australia.

Family VOMBATIDAE 2 or possibly 3 species

Common Wombat or Ursine Wombat *Vombatus ursinus* (formerly *Phascolomis*)
Vombatus is derived from wombat, a native name for this animal in New South Wales *ursus* (L) a bear *-inus* (L) suffix meaning like. Inhabiting Australia and Tasmania.

Soft-furred or Hairy-nosed Wombat *Lasiorhinus latifrons*
lasios (Gr) hairy, shaggy *rhis* (Gr) genitive *rhinos*, the nose *latus* (L) broad, wide *frons* (L) brow, forehead. Inhabiting the southern part of Australia.

Family MACROPODIDAE about 47 species
makros (Gr) long, large *pous* (Gr), genitive *podos*, a foot; alluding to their long and powerful hind legs.

Hare Wallaby *Lagorchestes leporoides*
lagōs (Gr) a hare *orchēstēs* (Gr) a dancer *lepus* (L) a hare *-oides*

(New L) from *eidos* (Gr) shape, form. The hare wallabies are hare-like in colour and habits; they live in the southern and western parts of Australia and some islands off the coast.

Banded Hare Wallaby   *Lagostrophus fasciatus*
lagōs (Gr) a hare   *strophis* (Gr) a band, a girdle   *fascia* (L) a band, a girdle   *-atus* (L) suffix meaning provided with; it has dark cross-bands down the back. Inhabiting south-western Australia.

Brush-tailed Rock Wallaby   *Petrogale penicillata*
petra (Gr) a rock   *galē* (Gr) a marten-cat or weasel   *penicillus* (L) a painter's brush   *-atus* (L) provided with. It lives amongst rocks and boulders and is widespread in Australia.

Ringtailed Rock Wallaby   *P. xanthopus*
xanthos (Gr) yellow   *pous* (Gr) a foot; 'yellow-footed'; it has dark rings along the tail and yellow marking on the limbs. Widespread in Australia in rocky habitat.

Little Rock Wallaby   *Peradorcus concinna*
pēra (Gr) a pouch, a wallet   *dorkas* (Gr) a gazelle or antelope; a reference to its agility among rocks, like the chamois   *concinnus* (L) pleasing proportion, harmony of form. Widespread in Australia in rocky habitat.

Nail-tail Wallaby   *Onychogalea frenata*
onux (Gr), genitive *onukhos*, a nail, a claw; the tip of the tail ends in a horny spur   *galē* (Gr) a marten-cat or weasel: *freno* (L) I bridle, I curb; this refers to the face marking which resembles a bridle. Inhabiting grassy plains in Australia.

Red-legged Pademelon   *Thylogale stigmatica*
thulakos (Gr) a bag, a pouch   *galē* (Gr) a marten-cat or weasel: *stigma* (L), genitive *stigmatis*, a brand put upon slaves   *-ica* (New L) a suffix sometimes used to emphasise a certain characteristic; an allusion to a mark like a brand on the neck: pademelon is from paddymalla, an old native name in Australia. Inhabiting eastern parts of Australia.

Red-necked Pademelon   *T. thetis*
Halmaturus thetis (now *Thylogale thetis*). 'M. F. Cuvier has named *Kangourou thetis* a new species, brought from Port Jackson, in 1826 by the frigate *La Thetis*, commanded by M. de Bougainville, and

which M. Busseuil is to describe in the narrative which is to be published of this voyage.' (R. P. Lesson, 1827.) This refers to Admiral H. Y. P. Baron de Bougainville (c. 1781–1846), son of the famous vice-Admiral L. A. Baron de Bougainville. This pademelon inhabits a coastal strip of eastern Australia.

Red-bellied Pademelon *T. billardieri*
Named by Professor Desmarest, of the Paris Museum, in 1822, in honour of the naturalist and navigator Jacques J. H. de La Billardière (1755–1834). He obtained the original Tasmanian specimen while on a voyage in search of Captain J. F. Comte de La Pérouse (1741–1788) who was lost off the New Hebrides in 1788. This pademelon inhabits the southern part of Australia, and Tasmania.

Bruijn's Pademelon *T. bruijni*
Named after the Dutch painter C. de Bruijn who provided the first exact description of a kangaroo in 1714. It was the first kangaroo known to Europeans. In 1711 it had been seen at Batavia (now Djakarta) in Java, in the Dutch Governor's garden, where he kept one as a pet or perhaps more as a curiosity. It must have been brought there by travellers as this pademelon lives in New Guinea.

Brush Wallaby *Wallabia rufogrisea*
Wallabia is derived from a native Australian word *wolobā* *rufus* (L) red, ruddy *griseus* (New L) derived from the German *greis*, grey. Sometimes known as the Red-necked Wallaby, it has a reddish-brown coat and grey chest and belly. It lives among brushwood and small trees in the eastern part of Australia, and Tasmania.

Pretty-faced Wallaby *W. elegans*
elegans (L) neat, elegant. Living in Queensland and possibly New South Wales, it is now very rare.

The Euro or Wallaroo *Macropus robustus*
makros (Gr) long, large *pous* (Gr) the foot *robustus* (L) strong. *Euro* and *walaru* are native names for this wallaby. It inhabits rocky areas of Australia.

Red Kangaroo *M. rufus*
rufus (L) red, ruddy. Found in most parts of Australia.

Great Grey Kangaroo *M. canguru*
canguru is a native name for the animal and is said to mean 'I don't know'. Inhabiting eastern Australia.

Quokka or Short-tailed Pademelon   *Setonix brachyurus*
seta (= *saeta*) (L) a bristle, stiff hair   *onux* (Gr) a claw   *brakhus* (Gr)
short   *oura* (Gr) the tail. *Quokka* is an Australian native name for a
pademelon. Inhabiting south-western Australia.

Lumholtz's Tree Kangaroo   *Dendrolagus lumholtzi*
dendron (Gr) a tree   *lagōs* (Gr) a hare; it is a tree-climber   Dr
C. S. Lumholtz (1851-1922) was a Norwegian zoologist and author
and was in Queensland during the years 1881 to 1893. These kangaroos live in Queensland.

Bennett's Tree Kangaroo   *D. bennettianus*
Named after Dr G. Bennett (1804-1893) a surgeon and zoologist
who spent most of his life in Australia. This kangaroo inhabits
Queensland and the island of New Guinea.

Goodfellow's Tree Kangaroo   *D. goodfellowi*
W. Goodfellow (1866-1953) was a zoologist who collected these
kangaroos in New Guinea during the years 1909 to 1911; he was
perhaps the greatest bird collector ever, chiefly for the British Museum
(Natural History). He made many expeditions from 1898 to 1925 and
is said to have brought back skins by the thousand. This kangaroo
lives in New Guinea.

New Guinea Forest Wallaby   *Dorcopsis muelleri*
dorkas (Gr) a gazelle   *opsis* (Gr) appearance; suggesting a small
kangaroo, in the same way that in Africa a gazelle is a small antelope,
wallabies being small kangaroos. Named after Salomon Müller
(1804-1863), a Dutch zoologist. He collected in Netherlands New
Guinea and worked with Professor H. Schlegel (1804-1884) o
Leiden Museum.

Short-nosed Rat Kangaroo or Bettong   *Bettongia cuniculus*
Bettong is a native name for the rat kangaroo   *cuniculus* (L) a rabbit
can also mean an underground passage. These small kangaroo
burrow in the ground and often live in rabbit warrens with th
rabbits that were brought to Australia.

Rufous Rat Kangaroo   *Aepyprymnus rufescens*
aipus (Gr) high   prumnos (Gr) the hindmost, endmost; alluding t
the long hind legs   *rufus* (L) red, ruddy   *-escens* (L) suffix meanin
beginning to, slightly; in this case 'reddish'. Inhabiting the Queens
land area of Australia.

Desert Rat Kangaroo *Caloprymnus campestris*
kalos (Gr) beautiful, fair *prumnos* (Gr) hindmost, endmost; an allusion to the well developed hind legs *campestris* (L) a plain, level country; it lives in the desert areas of central Australia.

Long-nosed Rat Kangaroo or Potoroo *Potorous tridactylus*
Potoroo is a native name in New South Wales for this small kangaroo *treis, tria* (Gr) three *daktulos* (Gr) a finger, a toe; this rat kangaroo has one toe absent and two others fused together, giving the appearance of three toes. Potoroos are now extinct in some parts of Australia but can still be found in Tasmania.

Musky Rat Kangaroo *Hypsiprymnodon moschatus*
hupsos (Gr) height, also the top, summit *prumnos* (Gr) the hindmost, endmost; alluding to the disproportionate development of the hind legs *odous* (Gr) genitive *odontos*, a tooth; the reason for 'tooth' is obscure, and not given in T. S. Palmer's standard work *Index Generum Mammalium* (1904) *moskhos* (Gr) musk and *moschatus* (New L) musky; it has musk glands that produce a strong odour. Inhabiting limited areas of Queensland.

8   Insect-eating Animals   INSECTIVORA

This is a strange assortment of rather primitive little animals, inhabiting almost the entire world, except Australia, the polar regions, and most of South America. They are to be found everywhere, not only in the country but in city parks and suburban areas, and yet very few people have actually seen a live shrew or mole, except possibly in captivity; hedgehogs are most often seen dead on the roadside after being hit by traffic.

They are generally of flesh eating habits, and in addition to insects they eat grubs, worms and snails; to obtain their prey some burrow in the earth, some hunt on the surface, and some, such as the water shrews and desmans, are good swimmers and catch their prey in the water.

Subclass   EUTHERIA

eu- (Gr) prefix meaning well, nicely; sometimes used to mean the typical animals in a group   *thēr* (Gr) a wild animal (see page 32 for notes about the placenta).

Order   INSECTIVORA

voro (L) I devour.

Family   SOLENODONTIDAE   2 species.

Cuban Solenodon   *Solenodon cubanus*
ōlēn (Gr), genitive *sōlēnos*, a channel, a pipe   *odous* (Gr), genitive

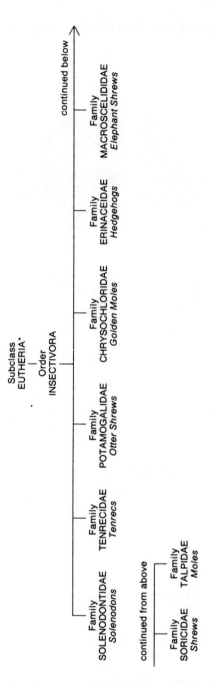

Subclass
EUTHERIA*
|
Order
INSECTIVORA

Family
SOLENODONTIDAE
Solenodons

Family
TENRECIDAE
Tenrecs

Family
POTAMOGALIDAE
Otter Shrews

Family
CHRYSOCHLORIDAE
Golden Moles

Family
ERINACEIDAE
Hedgehogs

Family
MACROSCELIDIDAE
Elephant Shrews

continued below

continued from above

Family
SORICIDAE
Shrews

Family
TALPIDAE
Moles

* Hereafter, all the described Orders are of the Subclass Eutheria and so only the Orders and Families are included in the tables accompanying Chapters 8 to 24.

odontos, a tooth, a fang; an allusion to the groove in the second incisor on each side of the lower jaw; it was originally supposed to be a channel for poisonous saliva, but tests have shown the poison to be extremely weak. Some now consider the groove may simply be to give strength to the tooth. -*anus* (L) suffix meaning belonging to; it lives in the Bayama Mountains in Cuba.

Haitian Solenodon *S. paradoxus*
para (Gr) contrary to, against *doxa* (Gr) opinion *paradoxos* (Gr) unexpected, strange. This species is found only in Haiti.

Family TENRECIDAE about 30 species
These animals are also known as tanrecs, or tendracs, from the Malagasy word *tandraka*; they are found only in Madagascar.

Common Tenrec *Tenrec ecaudatus*
ex (=*e*) (L) a prefix meaning from, out of; can mean without, like the Greek prefix *a*- *cauda* (L) the tail of an animal: -*atus* (L) suffix meaning provided with; 'not provided with a tail'; the common tenrec does not have a tail. It has sharp spines on the body rather like a hedgehog.

Hedgehog Tenrec *Setifer setosus*
saeta (=*seta*) (L) a bristle *fero* (L) I bear, I carry -*osus* (L) suffix meaning full of; 'a very bristly bristle carrier'.

Pygmy Hedgehog Tenrec *Echinops telfairi*
ekhinos (Gr) a hedgehog *ops*, from *opsis* (Gr) aspect, appearance: named after C. Telfair (1777–1833), a zoologist who founded the Botanical Gardens in Mauritius.

Banded or Streaked Tenrec *Hemicentetes semispinosus*
hēmi- (Gr) half, + *Centetes* (*Tenrec* formerly *Centetes*); indicating that this genus *Hemicentetes* differs from *Tenrec* in the presence of a third upper incisor, smaller canines, and the form of the skull *semi-* (L) half *spina* (L) a thorn, the prickles of animals -*osus* (L) suffix meaning full of; 'half-prickly'; this tenrec has bands of prickles alternating with bands of coarse black hair.

Rice Tenrec *Oryzorictes hova*
oruza (Gr) rice *oruktēs* (Gr) one who digs; they burrow in the rice fields *Hova* a dominant race in Madagascar, specially of the middle class and a name often used for animals inhabiting Madagascar.

Long-tailed Tenrec *Microgale longicaudata*
mikros (Gr) small *galē* (Gr) a marten-cat or weasel *longus* (L) long *cauda* (L) the tail of an animal *-atus* (L) suffix meaning provided with.

Web-footed or Water Tenrec *Limnogale mergulus*
limnē (Gr) a marshy lake *galē* (Gr) a marten-cat or weasel; it is a good swimmer, and one specimen was collected in a marsh *mergo* (L) I dip, I plunge *-ulus* (L) adjectival ending denoting tendency; the hind feet are fully webbed.

Family POTAMOGALIDAE 3 species
Giant Water Shrew or Giant Otter Shrew *Potamogale velox*
potamos (Gr) a river *galē* (Gr) a marten-cat or weasel; a good swimmer, it is a shrew, and not an otter or a marten-cat *velox* (L) fast, swift. Inhabiting western and central Africa.

Dwarf Otter Shrew *Micropotamogale lamottei*
mikros (Gr) small *potamogale* (see above) named after the French zoologist Dr M. Lamotte who collected in West Africa. This shrew inhabits the coastal regions of West Africa.

Family CHRYSOCHLORIDAE about 20 species
khrusos (Gr) gold *khlōros* (Gr) green, can mean honey-coloured or yellow; a reference to the beautiful iridescent hair.

Cape Golden Mole *Chrysochloris asiatica*
asiatica, of Asia. This is a misleading name as it does not belong to Asia. It was given this name by Linnaeus because it was mistakenly thought to come from Siberia; in fact it lives in South Africa.

De Winton's Golden Mole *Cryptochloris wintoni*
kruptos (Gr) secret, hidden *khlōros*, see above; they live a 'hidden' life, almost entirely underground. W. E. de Winton (1856–1922) was at one time Superintendent of the Zoological Gardens, London. This mole inhabits Africa south of the equator.

Hottentot Golden Mole *Amblysomus hottentotus*
amblus (Gr) blunt, point taken off *sōma* (Gr) the body; 'blunt-body'; it has brilliant iridescent golden fur and a bulb-shaped body. The Hottentots are a native people of Africa, and the name Hottentot is from the Dutch, meaning a stutterer, on account of their peculiar language. Range similar to above.

Family   ERINACEIDAE   about 15 species
erinaceus, or *ericius* (L) a hedgehog.

Moon Rat   *Echinosorex gymnurus*
ekhinos (Gr) a hedgehog   *sorex* (L) a shrew-mouse   *gumnos* (Gr)
naked   *oura* (Gr) the tail; the tail is naked and scaly. Inhabiting
southeast Asia.

Lesser Gymnure   *Hylomys suillus*
hulē (Gr) a wood, a forest   *mus* (Gr) a mouse: *sus* (L) a pig, hence
suillus (L) of swine; it looks rather like a small hog, hence 'hedgehog'.
Inhabiting Burma, Thailand, Borneo, Sumatra and South China.

Hedgehog   *Erinaceus europaeus*
erinaceus (L) a hedgehog   *europaeus* (L) belonging to Europe. This is
the common hedgehog of the British Isles and Europe. It also ranges
across parts of Asia.

Family   MACROSCELIDIDAE   14 species, possibly more.

Short-eared Elephant Shrew or Trumpet Rat   *Macroscelides
proboscideus*
makros (Gr) long   *skelis* (Gr), genitive skelidos, the leg; it has long
hind legs rather like the kangaroo   *proboscis* (L), genitive *proboscidis*,
a proboscis; derived from *pro* + *boskō* (Gr) I feed   *-ideus* (New L)
suffix meaning similarity. Inhabiting northern Africa.

Chequered Elephant Shrew   *Rhynchocyon cirnei*
rhunkhos (Gr) the snout, beak   *kuōn* (Gr) a dog; 'dog-nosed'; the
small proboscis is more like that of an elephant. Peters named this
shrew after Herr Cirne who stayed for two months in the Bororo
district of Mozambique, where this shrew was found. Inhabiting
central areas of Africa.

Family   SORICIDAE   more than 200 species; a large but uncertain number.
sorex (L), genitive *soricis*, the shrew-mouse.

European Pygmy Shrew   *Sorex minutus*
minutus (L) very small, minute. However, it is not the smallest shrew,
this honour probably belongs to Savi's Pygmy Shrew *Suncus etruscus*.

Common European Shrew *S. araneus*
aranea (L) a spider, hence *araneus*, relating to a spider; this is a reference to the old belief that both the shrew and the spider are poisonous.

Smoky Shrew *S. fumeus*
fumeus (L) smoky; the fur is greyish to brown on the back and a dirty white beneath. Inhabiting North America.

Pacific Water Shrew *S. bendirei*
Major C. E. Bendire (1836–1897) was an American zoologist and author. Inhabiting North America. A good swimmer.

Northern Water Shrew *S. palustris*
paluster (L), genitive *palustris*, marshy, boggy. Inhabiting Alaska and the northern part of North America.

North American Pygmy Shrew *Microsorex hoyi*
mikros (Gr) small *sorex* (L) the shrew-mouse: named after Dr Philip R. Hoy (1816–1892), an American explorer and naturalist. It is the smallest mammal in the New World and inhabits the northern part of North America.

Water Shrew *Neomys fodiens*
neō (Gr) I swim *mus* (Gr) a mouse *fodio* (L) I dig; they dig holes in the banks of rivers. Inhabiting Europe, including most of the British Isles, and also parts of Asia.

Mediterranean Water Shrew *N. anomalus*
anomalos (Gr) different, irregular; a reference to the absence of the keel of stiff hairs under the tail as possessed by *fodiens*. It inhabits not only Mediterranean areas, but can also be found in quite high mountains in Asia Minor.

Short-tailed Shrew *Blarina brevicauda*
Blarina is a coined name for certain shrews given by Gray in 1838 who was an inveterate 'coiner'; it has been suggested that it derives from Blair, Nebraska, and *-inus* (L) pertaining to; this shrew can be found in that part of North America, but it is now considered that this explanation is an invention, and is not valid *brevis* (L) short *cauda* (L) the tail of an animal. Inhabiting central parts of North America.

Least Shrew *Cryptotis parva*
kruptos (Gr) hidden *ous* (Gr), genitive *ōtos*, the ear; the ears are very

small and hidden under the fur   *parvus* (L) small; it is not so small as *Microsorex hoyi* (see above), but smaller than a mouse. Inhabiting North America.

Savi's Pygmy Shrew   *Suncus etruscus*
suncus from 'far sunki' (Arabic), a shrew   *etruscus*, an Etruscan, i.e. an inhabitant of Etruria, a district in north-western Italy, now known as Tuscany. Paolo Savi (1798–1871) was an Italian geologist and ornithologist and a Professor of Zoology at the University of Pisa. It inhabits large areas of southern Europe as well as Italy, and is considered to be the world's smallest mammal.

House Shrew   *S. murinus*
mus (L), genitive *muris*, a mouse   *murinus* mouselike. It has a vast range in continental Europe, central and southern Asia, and ranging to Indonesia.

Lesser White-toothed Shrew   *Crocidura suaveolens suaveolens*
krokus (Gr) nap, pile of cloth, a flock of wool   *oura* (Gr) the tail; the tail has short bristles interspersed with longer projecting hairs *suavis* (L) sweet   *olens* (L) smelling: *suave-olens* (L) sweet-smelling. Sometimes known as musk shrews, a reference to the musky odour which is less potent than that of common shrews. The white-toothed shrews have less red pigmentation on the teeth than other shrews; this shrew inhabits southern Europe, North Africa, and parts of Asia.

Scilly White-toothed Shrew   *C. s. cassiteridum*
kassiteros (Gr) tin   *kassiterides* (Gr) the tin islands, or Scilly Isles off the Cornish coast of England. There is no evidence that tin was found on the Scilly Isles but in the Early Iron Age tin was worked in Cornwall. This shrew was found on one of the islands comparatively recently, about 1924.

Armoured or Girder-backed Shrew   *Scutisorex congicus*
scutum (L) an oblong shield   *sorex* (L) the shrew-mouse   *-icus* (L) suffix meaning belonging to; *congicus*, of Congo, now Zaire. This little animal has the most remarkable backbone, resembling in miniature a massive girder designed to carry enormous weights; the reason for the elaborate structure is not known. This shrew is found only in Zaire.

Someren's Girder-backed Shrew   *S. somereni*
Dr R. A. L. Someren (1880–1955) was a naturalist who lived in

Uganda for many years from about 1905 onwards. This shrew was first discovered in Uganda.

Family   TALPIDAE   about 20 species
talpa (L) a mole.

Russian Desman   *Desmana moschata*
'Desman' is a Swedish word meaning musk   *moskhos* (Gr) musk; the desmans have glands under the tail which produces an overpowering musky smell. Inhabiting the central part of Russia.

Pyrenean Desman   *Galemys pyrenaicus*
galē (Gr) a marten-cat or polecat   *mus* (Gr) a mouse   *-icus* (L) suffix meaning belonging to; 'of the Pyrenees'. The desmans have webbed feet and are good swimmers. This one inhabits the northern part of Spain and Portugal in addition to the Pyrenees.

European Mole   *Talpa europaea*
Inhabiting Europe and parts of Asia.

Mediterranean Mole   *T. caeca*
caecus (L) blind; the moles, living underground, are usually almost completely blind. Inhabiting south-western parts of Europe.

Eastern Mole   *T. micrura*
mikros (Gr) small   *oura* (Gr) the tail; the very short tail is hidden under the fur. Inhabiting eastern Asia including Japan.

American Shrew Mole   *Neurotrichus gibbsi*
neos (Gr) new   *oura* (Gr) the tail   *thrix* (Gr), genitive *trichos*, the hair of man or beast; 'new hairy-tail'. George Gibbs (1815-1873) was a scientist from Oregon, U.S.A. This shrew mole is found on the west coast of North America from British Columbia to central California.

Townsend's Shrew Mole   *Scapanus townsendi*
scapanē (Gr) a spade or hoe; a reference to the claws, specially adapted for digging. J. K. Townsend (1809-1851) was an author who travelled and collected in the U.S.A. This mole lives in the northern coastal regions of North America.

Broad-footed Mole   *S. latimanus*
latus (L) broad   *manus* (L) the hand. Inhabiting the southern regions of North America.

Hairy-tailed Mole   *Parascalops breweri*
para (Gr) beside, near; i.e. resemblance   *skalops* (Gr) the digger, the
mole. Dr T. M. Brewer (1814–1880) was a zoologist and author of
Boston, Massachusetts. This mole lives in the eastern region of North
America and parts of northern Carolina.

Eastern American Mole   *Scalopus aquaticus*
skalops (Gr) the digger, the mole   *pous* (Gr) the foot; 'a foot for
digging'   *aqua* (L) water, hence *aquaticus* (L) living in water;
although this mole is a good swimmer it does *not* live in the water.
Inhabiting the south-eastern part of North America.

Star-nosed Mole   *Condylura cristata*
kondulos (Gr) knob of a joint, a knuckle   *oura* (Gr) the tail; a misleading name based on a faulty drawing by De La Faille   *crista* (L) the
crest of an animal or bird, hence *cristatus* (L) crested; it has a 'star' or
crest of 22 pink fleshy fingers growing on the snout. Inhabiting eastern
North America.

9 **Bats** CHIROPTERA

The bats are divided into two main groups; the suborder Megachiroptera, meaning 'large hand-wing', and the suborder Microchiroptera 'small hand-wing'. The scientific name of the order
Chiroptera is apt, as their wings are attached to their 'hands' and
their method of flight is peculiar in that they appear to 'claw their
way through the air'. Slow motion pictures show this clearly, and it
is quite different from the wing action of a bird. The group names
'large' and 'small' are not so apt, as some bats in the large group,
Megachiroptera are smaller than some of the big ones in the small
group Microchiroptera. However, the division is a good one having
regard to the structure of the skeleton and their eating habits.
Insect-eating bats of the suborder Microchiroptera have the remarkable ability to fly in the dark using ultra high frequency sound waves
to avoid obstacles and even track down tiny insects in flight and
catch them. This is all done without using their eyes and it is known
as echolocation. They emit these sound waves continuously during
flight, which are reflected by any object in the flight path and picked
up by the bat's ears. This is a quite different phenomenon from that
of cats, which are supposed to be able to see in the dark. In fact, cats
cannot see in complete darkness, although their eyes are sensitive to

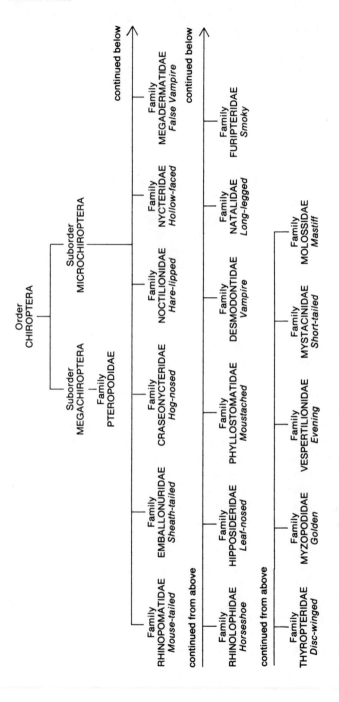

a very small amount of light, a characteristic of all animals that like to hunt at night. My own cat on one occasion, on a very dark winter's night, came in through the kitchen window and jumped down straight into a bucket of water! No self-respecting bat using echo-location would ever make such a ridiculous and humiliating mistake.

Subclass EUTHERIA (see pages 32 and 51)
Order CHIROPTERA
kheir (Gr) the hand *pteron* (Gr) wings (see introductory notes)

Suborder MEGACHIROPTERA fruit-eating bats
megas (Gr) big fruit-eating bats; 'big hand-wing'.

Family PTEROPODIDAE about 150 species
pteron (Gr) wings *pous* (Gr), genitive *podos*, the foot; 'wing-footed'; an allusion to the wing membrane which arises from the side of the back and the back of the second toe.

Indian Short-eared Fruit Bat *Cynopterus brachyotis*
kuōn (Gr), genitive *kunos*, a dog *pteron* (Gr) wings; 'a winged dog'; it has a dog-like head *brakhus* (Gr) short *ous* (Gr), genitive *ōtos*, the ear. Inhabiting Sri Lanka and south-east Asia, and ranging from southern China to Celebes.

Fruit Bat or Flying Fox *Pteropus vampyrus*
Pteropus, see above *vampir* (Slavonic). In eastern European folklore a vampire was a dead man who rose from the grave to prey upon the living; a 'blood-sucker'. The name is not appropriate as this bat is not a blood-sucker like the true vampire bats (see page 68). Inhabiting Burma and ranging east to Java, the Philippines, and Timor.

Grey-headed Flying Fox *P. poliocephalus*
polio (L) I polish; can mean I whiten, I adorn *kephalē* (Gr) the head. Inhabiting coastal and northern Australia.

Hammer-headed Bat *Hypsignathus monstrosus*
upsos (Gr) height *gnathos* (Gr) the jaw; an allusion to the 'deeply arched mouth': monstrum (L) a monster *-osus* (L) a suffix meaning full of, or intensive; 'very monstrous'; it is exceptionally hideous. Inhabiting forest areas from Uganda to West Africa.

Straw-coloured Fruit Bat *Eidolon helvum*
dolon (Gr) an image, a phantom; evidently an allusion to its move-

ments; *eidolon* is also the name for a small winged figure, human or combining human with animal elements, and found in Greek vase painting *helvus* (L) light bay, yellow. Widespread in Africa south of the Sahara.

Arabian Straw-coloured Fruit Bat   *E. sabaeum*
Sabaean (L) a native of Saba, in southern Arabia.

Queensland Blossom Bat   *Syconycteris australis*
sukon (Gr) a fig   *nukteris* (Gr) a bat; it is not known that this bat eats figs though many fruit bats do. Its main diet is the nectar of flowering trees, hence the name 'blossom bat'   *auster* (L) the south   *-alis* (L) suffix meaning pertaining to. This is one of the smallest bats in the suborder Megachiroptera, or 'big bats', being only about 75 mm (3 in) long. Inhabiting the Queensland area of Australia.

Suborder   MICROCHIROPTERA   insect-eating bats

mikros (Gr) small   *kheir* (Gr) the hand   *pteron* (Gr) wings; 'small hand-wing'; (see introductory notes on page 61).

Family   RHINOPOMATIDAE   4 species
rhis (Gr), genitive *rhinos*, the nose   *pōma* (Gr), genitive *pomatos*, a lid, a cover; these bats have a flange of skin connecting the ears, a small pad on the nose, and valvular nostrils opening through a narrow slit

Mouse-tailed Bat   *Rhinopoma microphyllum*
Rhinopoma, see above   *mikros* (Gr) small   *phullon* (Gr) a leaf; this could be interpreted as 'small leaf nose cover': they have a tail similar to that of a mouse. Ranging from West Africa to India, and possibly to Burma, Thailand and Sumatra.

Family   EMBALLONURIDAE   about 48 species
emballō (Gr) I put in   *oura* (Gr) the tail; the tail is partially sheathed in the tail membrane and emerges through the membrane to lie on the upper surface; it is used as a rudder when flying.

Sheath-tailed Bat   *Emballonura monticola*
Emballonura, see above   *mons* (L), genitive *montis*, a mountain   *col* (L) I inhabit; it lives in the mountains of Malaya, Sumatra, Borneo Java and Celebes.

Tomb Bat   *Taphozous perforatus*
taphos (Gr) a tomb   *zōos* (Gr) alive, living; 'one that lives in a tomb

great numbers of these bats were found in the tombs by the French expedition which collected them during investigations in Egypt at the beginning of the nineteenth century   *perforatus* (L) perforated; the interfemoral membrane is perforated to accommodate the tail. Inhabiting Egypt, Kenya, Arabia, and north-western India.

Pouch-bearing Bat   *T. saccolaimus*
sakkos (Gr) a coarse cloth of hair, a sack or bag   *laimos* (Gr) the throat; this tomb bat, as well as some others, has a small sac on the throat the purpose of which is not definitely established. Ranging from India and Sri Lanka to Borneo, Sumatra and Java.

Yellow-bellied Tomb Bat   *T. flaviventris*
flavus (L) yellow   *venter* (L), genitive *ventris*, the belly; the usual colour of bats is brownish but this one is black with pale yellow beneath. Inhabiting eastern Australia.

Family   CRASEONYCTERIDAE   1 species
A new genus and species created as recently as 1974 by John Edwards Hill of the British Museum (Natural History). As in all cases, the name for the family is constructed from the generic name.

Hog-nosed Bat   *Craseonycteris thonglongyai*
krasis (Gr), genitive *kraseōs*, a mixing, a blending   *nukteris* (Gr), genitive *nukteridos*, a bat; an allusion to this bat having composite characters. Named by J. Edwards Hill in honour of his friend the late Kitti Thonglongya (died 1974) who discovered the animal in Thailand.

Family   NOCTILIONIDAE   2 species
nox (L), genitive *noctis*, night; there is no such word in Latin as *noctilio* or *noctilionis*, although there is a Latin word *vespertilio* (genitive *vespertilionis*) from *vesper* (L) evening, and meaning 'an animal of the evening', a bat. *Noctilio* and *noctilionis* appear to be coined words from *nox* and supposed to mean 'an animal of the night'.

Hare-lipped or Fish-eating Bat   *Noctilio leporinus*
Noctilio, see above   *lepus* (L), genitive *leporis*, a hare   *-inus* (L) suffix meaning like; it has full, swollen-looking lips, the upper lip being divided by a fold of skin; it patrols the sea and fresh waters and catches fish in its claws. Inhabiting Mexico, Central and South America, and the West Indies.

Family   NYCTERIDAE   about 20 species
nukteris (Gr), genitive *nukteridos*, a bat.

Egyptian Hollow-faced Bat   *Nycteris thebaica*
Thebai (Gr) Thebes, an ancient city of Upper Egypt, surviving today
in the ruins of Luxor. Also known as Slit-faced Bats, this refers to a
groove on the face extending from the nose to between the eyes; there
is a deep hollow in the forehead. This bat inhabits Egypt and ranges
south to Angola.

Family   MEGADERMATIDAE   5 species
megas (Gr) large   *derma* (Gr), genitive *dermatos*, the skin of man or
animals; from the large wing and interfemoral membrane.

Australian False Vampire Bat   *Macroderma gigas*
makros (Gr) large   *derma* (Gr) the skin   Gigas was a giant. Known
as 'false vampire bat' because although carnivorous it is not a 'blood
drinker'. Inhabiting Australia.

Family   RHINOLOPHIDAE   75 species, possibly more
rhis (Gr), genitive *rhinos*, the nose   *lophos* (Gr) a crest; this refers to
the nose-leaf, part of which forms a sort of crest.

Greater Horseshoe-nosed Bat   *Rhinolophus ferrumequinum*
ferrum (L) iron, or can mean almost any instrument made of iron
equinum (L) relating to horses; a horseshoe; part of the nose-leaf is in
the form of a horseshoe. It inhabits Europe, North Africa, northern
India, China, and Japan.

Lesser Horseshoe-nosed Bat   *R. hipposideros*
hippos (Gr) a horse   *sidēros* (Gr) iron, or anything made of iron; 'a
horse-iron' or horseshoe. Range as *R. ferrumequinum*; both these bats
inhabit parts of the British Isles.

Family   HIPPOSIDERIDAE   about 57 species
see *hipposideros* above.

Large Malayan Leaf-nosed Bat   *Hipposideros diadema*
diadema (L) a royal headband, a decoration; a reference to the nose
leaves. These bats are related to the horseshoe-nosed bats, above
and have a horseshoe-shaped nose-leaf. Inhabiting Burma, Australia
and the Solomon Islands.

Flower-faced Bat   *Anthops ornatus*
anthos (Gr) a flower   *ops* (Gr) the eye, face   *ornatus* (L) decorated; a
reference to the nose-leaves. From the Solomon Islands.

Family   PHYLLOSTOMATIDAE   about 136 species
phullon (Gr) a leaf   *stoma* (Gr), genitive *stomatos*, the mouth; these
bats have a spear-shaped leaf on the nose just above the mouth, but
not on it.

Moustached or Leaf-lipped Bat   *Chilonycteris rubiginosa*
kheilos (Gr) the lip, brim; of animals the muzzle, beak   *nukteris* (Gr)
a bat   *rubigo* (L) rust   *-osus* (L) suffix meaning full of, prone to;
the lower lip has plate-like outgrowths and the body is rust-coloured.
Inhabiting Central America, northern South America, the West
Indies and common in Trinidad.

Naked-backed Bat   *Pteronotus davyi*
pteron (Gr) wings   *nōtos* (Gr) the back; the wing membrane is connected with the middle line of the back instead of the sides of the
body as in closely related species. Named after Dr John Davy (1790–
1868), brother of Sir Humphry Davy, 'well known for his physio-
logical papers'. The range is similar to *Chilonycteris*, above.

Leaf-chinned Bat   *Mormoops megalophylla*
mormō (Gr) a hideous monster, a hobgoblin   *ops*, from *opsis* (Gr)
aspect, appearance   *megas* (Gr), genitive *megalon*, big   *phullon* (Gr)
a leaf; 'a big-leafed monster'; the chin has leaf-like projections.
Usually found in caves in southern North America, Central America,
Venezuela, Trinidad and the West Indies.

Mexican Big-eared Bat   *Macrotus mexicanus*
makros (Gr) long   *ous* (Gr), genitive *ōtos*, the ear   *-anus* (L) suffix
meaning belonging to; of Mexico. It ranges south to Guatemala.

Spear-nosed Bat   *Phyllostomus hastatus*
phullon (Gr) a leaf   *stoma* (Gr) the mouth   *hastatus* (L) armed with
a spear; it has a spear-shaped leaf on the mouth. Inhabiting Central
and South America.

Jamaican Long-tongued Bat   *Monophyllus redmani*
monos (Gr) alone, single   *phullon* (Gr) a leaf; it has only one nose leaf.
Named by Dr Leach after Dom R. S. Redman, of Jamaica, from whom
specimen was obtained. Inhabiting the West Indies. Dr W. E. Leach

(1790-1836) was an author and zoologist at the British Museum (Natural History) from 1813 to 1821.

White-lined Bat   *Vampyrops vittatus*
vampir (Slavonic); in eastern European folklore a monster, 'a blood-sucker'   *ops*, from *opsis* (Gr) aspect, appearance; it is not a true vampire bat and does not feed on blood   *vittatus* (L) bound with a ribbon or can mean striped; it is coffee-coloured with white stripes along the back. Inhabiting Costa Rica and ranging eastwards to Venezuela.

Family   DESMODONTIDAE   3 species
desmos (Gr) a bundle   *odous* (Gr), genitive *odontos*, a tooth; this refers to the two large curved, cone-shaped incisors in the upper jaw, apparently pressed together (i.e. bundled) and occupying the entire space between the canines.

Common Vampire Bat   *Desmodus rotundus*
rotundus (L) round; the body is spherical in shape. These are the true vampire bats, the 'blood-suckers'. They are unable to feed on anything except blood which they obtain from live animals. The specially adapted teeth, extremely sharp, enable them to make a 'scooping' incision to obtain the blood, reputedly without waking a sleeping animal. Inhabiting Central and South America.

Hairy-legged Vampire Bat   *Diphylla ecaudata*
di- from *dis* (Gr) two   *phullon* (Gr) a leaf; they do not have true nose-leaves, but have two small pads on the snout   *e-* (=*ex*) (L) out; can mean without   *cauda* (L) the tail   *-atus* (L) provided with; 'not provided with a tail'; they are tailless. Inhabiting tropical areas of Central and South America.

Family   NATALIDAE   about 10 species (see below)

Long-legged Bat   *Natalus stramineus*
This bat has a floating glandular disc in the front part of the head known as the natalid organ. It is the only creature having such a process and its purpose has not been established. The term 'natalid organ' derives from its generic name *Natalus*, but the origin of the name remains obscure, and is not given by T. S. Palmer in his standard work *Index Generum Mammalium* (1904). *Stramineus* (L) made

of straw; sometimes known as Straw-coloured Bats, and also Funnel-eared Bats, referring to the unusual shape of the ears. Inhabiting Mexico, Central and South America, and the West Indies.

Family   FURIPTERIDAE   2 species
furo (L) I rage, I rave   *pteron* (Gr) wings; 'winged furies'. The Furies, or Erinyes, in Greek mythology, were hideous avenging deities. Aeschylus represents them as 'daughters of the night', and Sophocles as 'daughters of darkness'. Several bats have been named after similar disagreeable mythical characters.

Smoky Bat   *Furipterus horrens*
horreo (L) I bristle; it has a very bristly muzzle. The body colour is grey. Inhabiting Panama and Brazil.

Family   THYROPTERIDAE   2 species
thureos (Gr) a door-shaped shield   *pteron* (Gr) wings; this refers to a suction disc where the wing is attached to the wrist and another at the ankle. These discs enable the bat to attach itself to a smooth surface.

Disc-winged or Tricoloured Bat   *Thyroptera tricolor*
tres or *tria* (L) three   *color* (L) colour. Inhabiting Central and South America.

Family   MYZOPODIDAE   only 1 species
muzō (Gr) I suck in   *pous* (Gr), genitive *podos*, the foot; 'sucker-footed one'; it has suction discs similar to *Thyroptera* (see above).

Golden Bat   *Myzopoda aurita*
aurum (L) gold, colour of gold   *-itus* (L) suffix meaning provided with—but more likely *aurita* (L) long-eared; it does have unusually long ears. A very rare bat and found only in Madagascar.

Family   VESPERTILIONIDAE   about 280 species
vesper (L) evening   *vespertilio* (L), genitive *vespertilionis*, a bat; an animal of the evening. These bats are found in all temperate and tropical regions throughout the world including the British Isles.

Common or Brown Bat   *Myotis myotis*
mus (Gr) a mouse   *ous* (Gr), genitive *ōtos*, the ear; an allusion to the

large ears. Ranging from France to south-western Asia and to southern China.

Whiskered Bat   *M. mystacinus*
mustax (Gr), genitive *mustakos*, a moustache; this refers to the whiskers *-inus* (L) suffix meaning like. Ranging from Ireland to Japan, and India to Borneo.

Fishing Bat   *Pizonyx vivesi*
piezō (Gr) I squeeze   *onux* (Gr) a claw; an allusion to the toes and claws, which are compressed. The describer, Ménégaux, gives no clue as to the identity of Vives who apparently did not discover this bat. H. A. Ménégaux himself (1857-1937) was a French zoologist who was at the Paris Museum of Natural History from 1901 to 1926. It lives on the western side of North America.

Common Pipistrelle   *Pipistrellus pipistrellus*
pipistrello (Italian) a bat. This is the small bat seen in the evening in the British Isles. It is widespread and ranges from Ireland east to Japan, Kashmir and Taiwan.

American Western Pipistrelle   *P. hesperus*
hesperos (Gr) evening; can mean western, land of the setting sun. Inhabiting the western part of North America and Mexico.

American Eastern Pipistrelle   *P. subflavus*
sub- (L) under, beneath   *flavus* (L) yellow; it has a reddish-brown coat and pale underparts. Inhabiting eastern and central parts of North America and ranging south to Honduras.

Noctule   *Nyctalus noctula*
nux (Gr), genitive *nuktos*, night   *nuktalos* (= *nustalos*) (Gr) drowsy; said to be an allusion to its habit of coming out at dusk: *nox* (L), genitive *noctis*, night   *-ulus* (L) suffix denoting tendency; 'usually at night'. It has a wide range including Great Britain, Europe, Asia, and ranging eastwards to Burma and Malaya.

Leisler's Bat   *N. leisleri*
Dr L. P. A. Leisler (1771-1813) was a German scientist and author. This bat ranges from Ireland to south-western Asia as far as Punjab.

Serotine Bat   *Eptesicus serotinus*
Eptesicus is a coined word, derived from *eptēn* (Aorist tense of *petomai*) (Gr) I fly, and *oikos* (Gr) a house, and meaning 'a house-flyer'; they

are often seen flying near houses and roosting under the eves   *serotinus*
(L) late, backward; this name does not appear to have any meaning
with regard to the bat's development or behaviour; it does not come
out particularly late in the evening. It is quite a large bat, known in
the British Isles, and inhabits Eurasia and North Africa.

Club-footed Bat   *Tylonycteris pachypus*
tulē (Gr) a swelling or lump, a pad   *nukteris* (Gr) a bat   *pakhus* (Gr)
thick   *pous* (Gr) the foot; 'a pad-footed bat'; the under surface of the
base of the thumbs and the soles of the feet are expanded into fleshy
pads. Inhabiting southern Asia, Sumatra, Java, Borneo, the Philippines and Celebes.

Frosted Bat   *Vespertilio murinus*
vespertilio (L) a bat   *murinus* (L) mouse-like   the name 'frosted' refers
to the basic fur colour being obscured by intermingled white or white-
tipped hairs. Widespread in Europe and Asia, it has been reported in
both Great Britain and Japan.

Hinde's Bat   *Scotoecus hindei hindei*
skotos (Gr) darkness   *oikeō* (Gr) I dwell; 'dwelling in darkness'.
Named after Dr S. L. Hinde (1863–1931) who was in East Africa from
1896 to 1901. This bat is widespread in East Africa and has been
recorded in Somalia, Ethiopia, southern Sudan, Kenya, Uganda,
Zaire and Tanzania.

Falaba Bat   *S. h. falabae*
falabae, of Falaba; this does not refer to the town of Falaba in Sierra
Leone, West Africa, but to a ship named 'Falaba'. Dr J. C. Fox, who
collected this bat, was nearly lost when the ship was 'barbarously'
sunk by natives. The bat has been recorded in Nigeria and Cameroun,
but the distribution will probably prove to be more widespread in
Africa.

Butterfly Bat   *Glauconycteris poensis*
glaukos (Gr) bluish-green, silvery-grey; the fur is light grey at the tips
nukteris (Gr) a bat   *-ensis* (L) suffix meaning belonging to; *poensis*
refers to the Island of Fernando Po, off the coast of Cameroun, West
Africa; it is found in other areas of Africa. It has a butterfly-like flight.

Red Bat   *Lasiurus borealis*
lasios (Gr) hairy   *oura* (Gr) the tail; sometimes known as the Hairy-
tailed Bat   *boreas* (L) the north wind, hence *boreus* northern   *-alis* (L)

suffix meaning pertaining to. Inhabiting North and South America and the West Indies.

Hoary Bat   *L. cinereus*
cinerea (L) ash-coloured; hoary can mean white, or greyish-white. The range is similar to *L. borealis*.

Barbastelle   *Barbastella barbastellus*
barbastelle (Fr) a little beard; they are sometimes known as Hairy-lipped Bats. Inhabiting Europe, north-eastern Africa and Asia.

European Long-eared Bat   *Plecotus auritus*
plekō (Gr) I twine, I twist   *ous* (Gr), genitive *ōtos*, the ear; referring to the unusual form of the junction at the base of the ears   *auris* (L) the ear, hence *auritus* (L) long-eared. Found in Europe, Asia and North Africa.

American Long-eared Bat   *P. macrotis*
makros (Gr) long   *ous* (Gr), genitive *ōtos*, the ear. Living in the south-eastern part of North America.

Spotted Bat   *Euderma maculata*
eu- (Gr) prefix meaning well, nicely but sometimes used to mean typical   *derma* (Gr) skin; in this case the interpretation would be 'nicely coloured skin'   *macula* (L) a spot   *-atus* (L) suffix meaning provided with; spotted. It has a unique colour pattern with white spots. A rare bat and found only in certain localities in western U.S.A. and north-western Mexico. Also known as the Death's Head Bat.

Tube-nosed Bat   *Murina suilla*
mus (L), genitive *muris*, a mouse, hence *murinus* mouse-like   *sus* (L), genitive *suis*, a pig and so *suillus* pig-like; the nostrils are at the end of tubes. Inhabiting south-east Asia, Sumatra, Borneo, the Philippines and other islands in that area.

Little Tube-nosed Bat   *M. aurata*
aurum (L) gold, hence *auratus* (L) golden, gilded; this refers to the golden-brown coat. Inhabiting south-eastern Siberia, Japan, Burma and northern India.

Painted Bat   *Kerivoula picta*
Kehelvoulha is a Singalese name for this bat   *pingo* (L) I paint and *pictus*, painted; the brightest of all bats, it has long woolly hair and the colour is orange or scarlet with black wings, and said to look like

a leaf in autumn. Inhabiting India, Sri Lanka, southern China, and Indonesia ranging east to Bali Island.

Pale or Desert Bat *Antrozous pallidus*

antron (Gr) a cave *zōon* (Gr) an animal; *zōos* alive, living; it usually roosts in caves *pallidus* (L) pale. Found only in the west of North America.

Family MYSTACINIDAE 1 species

mustax (Gr), genitive *mustakos*, a moustache; can mean the upper lip *inus* (L) suffix meaning pertaining to.

Short-tailed Bat *Mystacina tuberculata tuberculata*

tuber (L) a knob and so *tuberculum* (L) a small knob *-atus* (L) suffix meaning provided with; it has small knobs, or pimples, on the upper lip. Widespread in the North Island of New Zealand, and the northern area of South Island, and on the Barrier Islands. It is becoming very rare and possibly is now extinct.

Short-tailed Bat *M. t. robusta*

robustus (L) strong, robust. Inhabiting Stewart Island off the south coast of South Island, New Zealand.

Family MOLOSSIDAE 88 species, possibly more

molossos (Gr) a kind of wolf-dog used by shepherds; these bats have dog-like faces.

Celebes Naked or Hairless Bat *Cheiromeles parvidens*

kheir (Gr), genitive *kheiros*, the hand *melos* (Gr) a limb; 'a hand limb'; T. S. Palmer, in his standard work *Index Generum Mammalium* (1904) says 'Possibly in allusion to the first toe, which is separated from the others like a thumb and probably opposable to them, thus giving the foot the appearance of a hand' *parvus* (L) small *dens* (L) a tooth. Probably the most unattractive of the bats in appearance, and furthermore possessing glands that can produce a quite revolting smell. Inhabiting Celebes and the Philippines.

Philippine Naked or Hairless Bat *C. torquatus*

torquatus (L) having a collar; the skin round the neck is in folds, giving the appearance of a collar. This bat and *C. parvidens* differ only in small details. Inhabiting the Malay States, Indonesia and the Philippines.

Free-tailed Bat   *Tadarida taeniotis*
Tadarida is a dubious coinage by C. S. Rafinesque (1783-1840) a
scholar of strange origin who died in poverty, though later he was re-
interred in Lexington, Kentucky, U.S.A., where he had previously
taught botany at the University. He outlined a pre-Darwinian theory
of evolution. *taenia* (L) a head-band   *ous* (Gr), genitive *ōtos*, the ear;
'ear-bands'. The name 'free-tailed' refers to the tail being independent
of the posterior portion of the flight membrane. Inhabiting southern
Europe and ranging to eastern Siberia; also north-western Africa,
Egypt, Iran, Japan and Taiwan.

Railer Bat   *T. thersites*
Named after Thersites 'The ugliest man in the Greek camp before
Troy'; a relative of Diomedes, he was a railing demagogue, and this
bat is ugly and chatters constantly. It is a far-fetched and interesting
connection, but considered authentic. Inhabiting West Africa, Zaire,
Rwanda and possibly Mozambique and Zanzibar.

These rather extraordinary animals are somewhat of a puzzle for zoologists. They have no close relatives, and so have been given an order all to themselves. At one time they were thought to be bats, and at another time primates, but their anatomy shows that in neither case was this correct. One remarkable feature is the structure of the lower front teeth, formed like small combs, each having from 6 to 11 teeth, probably to facilitate leaf-eating and fur-cleaning. They have a membrane of skin extending from the forelegs to the hind legs and out to the tail, used when gliding; they cannot fly, in a true sense, but are probably the most expert of all mammals that are able to glide.

Subclass EUTHERIA (see pages 32 and 51)

Order DERMOPTERA

erma (Gr) the skin, hide of animals *pteron* (Gr) a feather; can mean wings.

Family CYNOCEPHALIDAE 2 species
(formerly GALEOPITHECIDAE)
kuōn (Gr), genitive *kunos*, a dog *kephalē* (Gr) the head; they have a somewhat dog-shaped head; they are not lemurs in spite of the name.

Order
DERMOPTERA
|
Family
CYNOCEPHALIDAE
Flying Lemurs

Malayan Flying Lemur   *Cynocephalus variegatus*
(formerly *Galeopithecus*)
varius (L) variegated; the coat is a mottled grey, fawn and buff.
Inhabiting south-east Asia, including Sumatra, Java and Borneo.

Philippine Colugo or Flying Lemur   *C. volans*
volo (L) I fly and *volans* flying. Sometimes called the Cobego; inhabiting the Philippines.

11 Tree Shrews, Lemurs, Monkeys, Apes and Man PRIMATES

This large group consists of the Apes, the Monkeys and Humans, but it also includes a number of less monkey-like creatures such as the Lorises and the Lemurs. So the Order is divided into two Suborders; Lemuroidea the 'lemur-like', and Anthropoidea the 'man-like'.

Then there are the Tree-shrews, anatomically quite a problem, but zoologists do not all agree that these should be included with the primates. There is a tendency now to revert to their previous classification as insectivores. Scientific research continues, so until some agreement is reached they can remain with the primates.

The reader will probably experience some difficulty regarding the English names. For example, langurs, baboons, gibbons, orang-utans, gorillas and chimpanzees, to mention only a few; how do we know which of these are monkeys and which are apes? As a general rule it can be said that monkeys have tails and apes do not. Thus, the apes are gibbons, orang-utans, gorillas and chimpanzees.

With the exception of the human race, and a few monkeys that reach Europe and the north-eastern parts of Asia, the primates live in tropical countries.

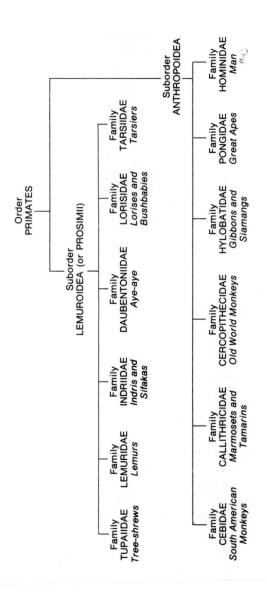

Subclass   EUTHERIA   (see pages 32 and 51)

Order   PRIMATES
primus (L) first, foremost

Suborder   LEMUROIDEA (or PROSIMII)
lemures (L) ghosts, spectres; in ancient Rome the spirits of the evil dead; supposed to be so named because of the animal's nocturnal habits and stealthy movements   *-oides* (New L) from *eidos* (Gr) apparent shape, form, a kind, sort.

Family   TUPAIIDAE   about 20 species
tupai is a Malayan name that is used for various small squirrel-like creatures.

Common Tree Shrew   *Tupaia glis*
Tupaia see above   *glis* (L) a dormouse; it is not a dormouse though rather similar. Inhabiting India, Burma, China and the East Indies.

Madras Tree Shrew   *Anathana ellioti*
anatheō (Gr) I run up   Dr D. G. Elliot (1835–1915) was an American zoologist, at one time Curator of the Field Museum of Natural History, Chicago. Inhabiting southern Asia.

Smooth-tailed Tree Shrew   *Dendrogale murina*
dendron (Gr) a tree   *galē* (Gr) a marten-cat or weasel; the word *galē* has been used in nomenclature for various small mammals   *mus* (L), genitive *muris*, a mouse, hence *murinus* (L) mouse-like, but they are bigger than a mouse. Inhabiting Vietnam and the Kmer Republic.

Mindanao Tree Shrew   *Urogale everetti*
ura (Gr) the tail   *galē* (Gr) see above; the tail is a distinctive feature as it is close-haired and not bushy as in the genus *Tupaia*. A. H. Everett (1848–1898) was a zoologist resident in Sarawak from 1869 to 1890. Mindanao is one of the Philippine Islands where these tree shrews live.

Pen-tailed Tree Shrew   *Ptilocercus lowi*
ptilon (Gr) a wing   *kerkos* (Gr) the tail of an animal; 'wing-tailed'; the tail has feather-like fringes of long white hairs. Sir Hugh Low (1824–1905) was Resident in Perak, Malaya, from 1877 to 1889. This tree shrew inhabits Borneo and the Malayan Peninsula.

Family   LEMURIDAE   about 15 species
lemures (L) ghosts, spectres, (see page 79). Lemurs are found only in Madagascar, now known as the Malagasy Republic.

Grey Gentle Lemur   *Hapalemur griseus*
hapalos (Gr) soft, gentle   *lemures* (L) see above; an allusion to the long soft fur   *griseus* (New L) grey, derived from the German greis.

Broad-nosed Gentle Lemur   *H. simus*
simus (L) flat-nosed, snub-nosed. This species now very rare.

Ring-tailed Lemur   *Lemur catta*
catta (New L) a cat. Probably the best known of the lemurs, it has striking black and white rings round the tail.

Ruffed Lemur   *L. variegatus*
variegatus (L) variegated; it usually has a black and white pattern on the body and a ruff of long yellowish hair around the neck, but the colour variation is quite remarkable.

Black Lemur   *L. macaco*
macacus (New L) derived from *macaco* (Port) a native name for a monkey.

Weasel or Sportive Lemur   *Lepilemur mustelinus*
lepos (L) pleasantness, charm   *mustela* (L) a weasel and so *mustelinu.* (L) of a weasel; 'weasel-like'.

Hairy-eared Dwarf Lemur   *Cheirogaleus trichotis*
kheir (Gr) the hand   *galē* (Gr) a marten-cat or weasel; an allusion to the long fingers and freely movable thumb, adapted for prehension *thrix* (Gr), genitive *trikhos*, the hair of both man and animals   *ou* (Gr), genitive *ōtos*, the ear. This lemur is now very rare.

Greater Dwarf Lemur   *C. major*
major (=*maior*) (L) greater.

Fat-tailed Dwarf Lemur   *C. medius*
medius (L) middle, intermediate. Now very rare.

Lesser Mouse Lemur   *Microcebus murinus*
mikros (Gr) small   *kēbos* (Gr) a monkey   *mus* (L), genitive *muris*, mouse, hence *murinus*, mouselike; probably the smallest living primate and not much bigger than a large mouse.

Fork-crowned Lemur   *Phaner furcifer*
phaneros (Gr) visible, evident   *furca* (L) a fork   *fero* (L) I bear, I carry;
it has black fork-shaped markings on the head. Now very rare.

Family   INDRIIDAE   4 species; subspecies are not included in the
count of the number of species in a family
The name 'indri' is said to originate from a Malagasy word meaning
'look' or 'there it is'; indris and sifakas are found only in Madagascar.

Woolly Indri   *Avahi laniger*
Avahi is a native name for the woolly lemur   *lana* (L) wool   *gero* (L)
I carry, hence *laniger* (L) wool-bearing. Like the lemurs, indris live
in Madagascar and are closely related.

Diadeemed Sifaka   *Propithecus diadema diadema*
pro (Gr) before, in front of; this probably refers to its relative age in
evolution, 'before apes'   *pithēkos* (Gr) an ape   *diadema* (L) a royal
headband; it has a headband of white fur   *sifac* is a Malagasy word.
The nominate subspecies (see pages 18 and 19).

Perrier's Sifaka   *P. d. perrieri*
M. Perrier de la Bathie was a French botanist in Madagascar; the
name was given quite recently, in 1931. This is a subspecies (see pages
18 and 19).

Black Sifaka   *P. d. edwardsi*
Professor Alphonse Milne-Edwards (1835–1900) was a French zoologist; in collaboration with another French zoologist, A. Grandidier
(1836–1921) he wrote a classical work on what was then Madagascar,
and its ecology; a subspecies (see pages 18 and 19).

Verreaux's Sifaka   *P. verreauxi verreauxi*
Named after J. B. E. Verreaux (1810–1868) a French zoologist.
Sifakas are found only in Madagascar where they live in the forests
round the coast. This sifaka, the nominate subspecies (see pages 13
and 19) is now very rare.

Coquerel's Sifaka   *P. v. coquereli*
Named after Dr C. Coquerel (1822–1867), a French zoologist who
lived for a time in Madagascar; a subspecies (see pages 18 and 19)
and now very rare.

Crowned Sifaka   *P. v. coronatus*
coronatus (L) provided with a crown. A rare subspecies (see pages 18 and 19).

Van der Decken's Sifaka   *P. v. deckeni*
Baron Van der Decken (1833–1865) was a Dutch explorer and naturalist. A rare subspecies (see pages oo and oo).

Indri   *Indri indri*
Indri, see above, under Family. Sometimes known as the Endrina, it lives in the tree-tops of the forests of eastern Madagascar; it is very rare.

Family   DAUBENTONIIDAE   1 species
Named after Louis J. M. Daubenton (1716–1799), the French zoologist, who discovered this animal in the year 1780.

Aye-aye   *Daubentonia madagascariensis* (*Daubentonia* formerly *Cheiromys*)
This peculiar animal presented a problem for the zoologists, as the anatomical features were most unusual, although they showed it to be a primate. It has an extraordinary long, thin third finger, probably used for digging the larvae of wood-boring insects out of the holes in the wood, and the teeth are similar to those of a rodent. It was first described as a squirrel, and was given eight different generic name between 1795 and 1846. It has now been given a family to occupy or its own; it is very rare. The name aye-aye is derived from a Malagasy name *aiay*; it inhabits eastern Madagascar.

Family   LORISIDAE   11 species
The name loris is derived from the Dutch *loeris*, meaning a clown, a booby; the lorises have a comical appearance.

Slender Loris   *Loris tardigradus*
tardus (L) slow   *gradus* (L) a step, a pace; all lorises take slow, deliberate steps. This loris lives in southern India and Sri Lanka.

Slow Loris   *Nycticebus coucang*
nux (Gr), genitive *nuktos*, night   *kēbos* (Gr) a monkey; 'night monkey'; they are nocturnal and the very large eyes enable them to see in a poor light: *kukang* is the native name in Malaya. The range from Assam to Tongking, south to Singapore, Sumatra, Java and Borneo.

Lesser Slow Loris   *N. pygmaeus*
pugmē (Gr) a fist and so *pugmaios* (Gr) about a foot long, or tall, i.e.
'dwarfish'. It lives in Vietnam.

Angwantibo   *Arctocebus calabarensis*
arktos (Gr) a bear   *kēbos* (Gr) a monkey   Calabar is in Nigeria
-ensis (L) suffix meaning belonging to   *angwantibo* is an Efik name for
this animal, the language of a people of south-eastern Nigeria; it
inhabits central West Africa.

Potto   *Perodicticus potto*
pēros (Gr) maimed, disabled   *deiktikos* (Gr) showing, proving; 'showing maimed'; this refers to the second fingers and toes, which are very
short, and look like stumps after an amputation; potto is a word of
Niger-Congo origin meaning a tailless monkey; this loris has a very
short tail. It lives in the forests of East, Central and West Africa.

Thick-tailed Bushbaby   *Galago crassicaudatus*
Galago is probably from *golokh* (Wolof) a monkey; the Wolof are a
people of the western Sudan   *crassus* (L) thick   *cauda* (L) the tail of
an animal   *-atus* (L) suffix meaning provided with. The range is
Gambia and surrounding areas in West Africa, and large areas in
East Africa down to Rhodesia.

Senegal Bushbaby   *G. senegalensis*
ensis (L) suffix meaning belonging to; the type species came from
Senegal but it inhabits other large areas of Africa.

Allen's Bushbaby or Moholi Galago   *G. alleni*
Rear-Admiral W. Allen F.R.S. (1793-1864) was an English zoologist
who led an expedition to Niger in 1841   *moholi*, or *maholi*, is an African
native name. Inhabiting large areas of Africa, from Kenya in the east
to Gambia in the west.

Demidoff's Bushbaby   *G. demidovi*
Paul Demidov (1738-1821) was a Russian scientist and traveller;
his name first appeared in a Moscow publication in 1808. This bushbaby lives in a small area in West Africa, mostly confined to Cameroun
and Gabon.

Family   TARSIIDAE   3 species

Philippine Tarsier   *Tarsius syrichta*
tarsos (Gr) the flat of the foot, the part between the toes and the heel;

an allusion to the tarsal bones of the hind legs which are exceptionally long   *suriktas* (Gr) a player on Pan's pipe, to make a whistling sound; tarsiers have a very high-pitched call. The Latin name was given by Linnaeus in 1758, more than 200 years ago! Inhabiting the Philippines.

Horsfield's or Western Tarsier   *T. bancanus*
Banka is an island in the East Indies lying just off the coast of Sumatra -*anus* (L) suffix meaning belonging to. Dr T. Horsfield (1773-1859) was a scientist who was in Sumatra from 1796 to 1818; he named many mammals and birds of the East. This tarsier lives in Sumatra and Borneo, but was found first on Banka Island.

Celebes or Eastern Tarsier   *T. spectrum*
spectrum (L) an image, an apparition; tarsiers are nocturnal and have very large eyes, which would suggest 'an apparition'. Celebes is a large island in Indonesia lying to the east of Borneo.

Suborder   ANTHROPOIDEA (or SIMIAE, PITHECOIDEA)
anthrōpos (Gr) man   -*oides* (New L) from *eidos* (Gr) shape, resemblance; 'man-like'.

Family   CEBIDAE   about 35 species
kēbos (Gr) a monkey.

Douroucouli or Night Monkey   *Aotus trivirgatus*
(formerly *Nyctipithecus*)
a- (Gr) prefix meaning not, or there is not   *ous* (Gr), genitive *ōtos* the ear; they have very small ears hidden under the fur   *tria* (L) three *virga* (L) a twig   -*atus* (L) suffix meaning provided with, so *virgatu* (L) can mean striped; they have black and brown lines on the face Douroucouli is a South American native name; it has also been known as the Owl Monkey on account of its 'owl-like' face, and 'nigh monkey' because it is active at night. Inhabiting central South America.

Collared Titi or Widow Monkey   *Callicebus torquatus*
kallos (Gr) a beauty; *kalos* (Gr) beautiful   *kebos* (Gr) a monke *torquatus* (L) wearing a collar or necklace; it has a collar of white fu Possibly named 'widow' because the coat is dark-coloured or eve black. Inhabiting a large area north of the Amazon in South Americ

Orabassu Titi   *C. moloch*

Moloch was a Semitic god to whom children were sacrified; the name probably has no significance (see notes about Linnaeus under Common Marmoset, page 87), but may be associated with its ugly face. Orabassu is derived from the Tupi native name *oyapussa*; *titi* is a Spanish South American name meaning 'little cat'. Inhabiting parts of the Amazon basin in South America.

Bald Uakari   *Cacajao calvus*   (formerly *Brachyurus*)

cacajao (Tupi) a South American native name   *calvus* (L) bald; the face and head of this monkey are almost hairless. It lives in the forests of the upper Amazon Basin area, South America.

Black-headed Uakari   *C. melanocephalus*

melas (Gr), genitive *melanos*, black, dark   *kephalē* (Gr) the head. Inhabiting the upper Amazon Basin. Uakari is a Tupi South American native name.

Hairy or Monk Saki   *Pithecia monachus*

pithēkos (Gr) an ape   *monakhos* (Gr) solitary, a monk; it has a cape of long hair on the head and shoulders resembling a monk's habit. Inhabiting the Guyana area and the Amazon Basin.

Pale-headed or White-faced Saki   *P. pithecia*

Saki is a name derived from a Tupi South American word. This saki has a remarkable white face; it inhabits the Guyana area and part of the northern Amazon Basin.

Black Saki   *Chiropotes satanus*

cheir (Gr) the hand   *potēs* (Gr) a drink; it drinks from the back of its hand by soaking the fur in water and licking it   *Satanas* (Gr) Satan, a Hebrew word meaning the enemy; it is sometimes used to indicate a black or an ugly animal. Inhabiting the Amazon area.

White-nosed Saki   *C. albinasus*

albus (L) white   *nasus* (L) the nose. Inhabiting a southern part of the Amazon Basin.

Guatemalan Howler Monkey   *Alouatta villosa*
(*Alouatta* formerly *Mycetes*)

Alouatta (New L) derived from the French alouate   *villus* (L) shaggy hair   *-osus* (L) suffix meaning full of, prone to; 'a lot of shaggy hair'. Inhabiting Central America.

Mantled or Panamanian Howler Monkey   *A. palliata*
pallium (L) a type of Greek mantle, hence *palliata*, clad in a pallium; the hair is longer on the sides and of a different colour giving the impression of a cloak. Howler monkeys have a special formation of the throat, a kind of sound-box, which increases the power of the voice; they live in southern Mexico and the northern part of South America.

Red Howler Monkey   *A. seniculus*
senex (L) old, aged; *seniculus* (L) a little old man; the name is an allusion to its appearance. Inhabiting South America.

Brown Howler Monkey   *A. fuscus*
fuscus (L) dark, dusky. Inhabiting south-eastern Brazil.

Red-handed Howler Monkey   *A. belzebul*
belzebul, another spelling of Beelzebub, god of flies, from the Hebrew *baal*, a lord, and *zeebub*, a fly; suggesting black, evil, an allusion to the rather ugly face. It inhabits a large area south of the Amazon.

Black Howler Monkey   *A. caraya*
The Caraya are a people living in the central part of South America where this monkey is found.

White-throated Capuchin   *Cebus capucinus*
kēbos (Gr) a monkey   the Franciscan Monks were known as Capuchins and the dark patch of hair on the crown of this monkey's head resembles a monk's cap. Inhabiting Central America and the north-western tip of South America.

Weeper or Black-capped Capuchin   *C. nigrivittatus*
niger (L) black   *vitta* (L) a ribbon, a band   *-atus* (L) suffix meaning provided with; 'having a black band'; they communicate by making a continuous 'weeping cry'. Inhabiting the Guyana and Surinam area and the northern Amazon Basin; there are other species of capuchin monkeys all living in South America; they are often sold as pets, and have been used by organ-grinders.

Common Squirrel Monkey   *Saimiri sciureus*
(formerly *Chrysothrix*)
Saimiri is from a Brazilian–Portuguese word for a small monkey *sciurus* (L) a squirrel; see notes about the suborder Sciuromorpha on page 113. Living in the forests of the Orinoco, Guyana and the Amazon area.

Black Spider Monkey *Ateles paniscus*
a- (Gr) a prefix meaning not, or there is not *teleios* (Gr) complete, entire; 'not complete'; it usually has no thumb though sometimes it is vestigial *paniscus* (L) a sylvan deity, 'a little pan', from Pan and the diminutive suffix *-iscus* (L). Inhabiting two large areas in the Amazon Basin.

Woolly Spider Monkey *Brachyteles arachnoides*
brakhus (Gr) short *teleios* (Gr) complete, entire; it has very short thumbs or none at all *arakhnēs* (Gr) a spider *-oides* (New L) derived from the Greek *eidos*, apparent shape, form. Inhabiting the forest area to the south-east of Brazil.

Humboldt's Woolly Monkey *Lagothrix lagothricha*
lagos (Gr) a hare *thrix* (Gr), genitive *trikhos*, hair of man or beast, wool: an allusion to the woolly, hare-like fur: Baron v. Humboldt (1769-1859) was a German scientist and explorer who lived in South America from 1799 to 1804. The coat of these monkeys is thick and woolly; they inhabit the north-western part of South America.

Peruvian Mountain Woolly Monkey *L. flavicauda*
flavus (L) yellow *cauda* (L) the tail of an animal; it has a yellow streak on the underside of the tail. This is a very rare monkey found only in the mountains of northern Peru.

Family CALLITHRICIDAE (formerly HAPALIDAE)
about 35 species
callos (Gr) a beauty *kalos* (Gr) beautiful *thrix* (Gr), genitive *trikhos*, hair; they have elaborate 'hair styles' and colourful coats.

Common Marmoset *Callithrix jacchus*
Callithrix formerly *Hapale*)
Iacchus (= Iacchus) a Roman god associated with Bacchus, the god of wine. In 1758 Linnaeus used several names from classical mythology for the specific names of certain species, and it is probable that they were given without thought of any physical significance. Marmoset is from *marmouset* (Fr) a grotesque figure, a young monkey (a little boy); it may have derived from *marmor* (L) marble, and later *marmoretum*, a little marble figure. This marmoset inhabits the eastern part of Brazil.

White-eared Marmoset   *C. aurita*
auris (L) the ear and hence *auritus* (L) long-eared. Inhabiting the eastern part of Brazil.

Silvery Marmoset   *C. argentata*
argentum (L) silvery and hence *argentatus* (L) ornamented with silver. Inhabiting central parts of South America south of the Amazon.

Pygmy Marmoset   *Cebuella pygmaea*
kēbos (Gr) a monkey   *-ellus* (L) diminutive suffix; 'a small monkey'; it is the smallest of the Anthropoidea   *pugmaios* (Gr) dwarfish, from *pugmē* (Gr) a fist. Inhabiting the northern part of Peru.

Golden Lion Marmoset   *Leontideus rosalia*
leo (L), genitive *leontos*, a lion   *-ideus* (New L) from *eidos* (Gr) the shape, resemblance   *rosa* (L) a rose   *-alis* (L) suffix meaning like: referring to the colour, a reddish gold. It inhabits the south-eastern area of Brazil.

Cotton-headed Tamarin   *Saguinus oedipus*   (formerly *Tamarin*)
sagouin (Fr) a squirrel-monkey   *-inus* (L) suffix meaning like   *oidipous* (Gr) swollen-footed; possibly no direct connection with this meaning. There is a legend about King Polybus whose shepherd found a baby abandoned on Mount Cithaeron, with injuries to his feet. The King reared him as his own son and named him Oedipus, because of his swollen feet. Inhabiting the north-western part of South America.

Pied Tamarin   *S. bicolor*
bicolor (L) of two colours; it is brown and white. Tamarin is a Caribbean word meaning a squirrel-monkey; this tamarin inhabits an area north of the Amazon.

Red-handed Tamarin   *S. midas*
Possibly an allusion to Midas, the legendary King of Phrygia, who was supposed to have had 'ass's ears'; this tamarin has unusually large ears. Inhabiting Brazil in an area around the mouth of the Amazon.

Brown-headed Tamarin   *S. fuscicollis*
fuscus (L) dark-coloured   *collum* (L) the neck. Inhabiting an area north of the Amazon.

Emperor Tamarin   *S. imperator*
imperator (L) a leader, a chief. A taxidermist, so the story goes, had never seen a live tamarin and twisted the white 'moustache' upward

to look like the Emperor of Germany, instead of letting it droop in the natural position. It thus acquired the name Emperor Tamarin as a joke, but the name stuck, and the Latin name became established as *Saguinus imperator*. Inhabiting the central part of South America.

Family   CERCOPITHECIDAE   about 60 species
kerkos (Gr) the tail of an animal   *pithēkos* (Gr) an ape. In fact this is a family of monkeys, not apes (see page 77).

Toque Monkey   *Macaca sinica*
Macaca is derived from *macaque* (Fr) from Portuguese, a monkey *sinica* is a coined word meant to mean 'of China'; it was given by Linnaeus in 1771 because he thought this monkey came from China. In fact, it inhabits Sri Lanka. It has a whorl of hair on the head.

Bonnet Monkey   *M. radiata*
radiatus (L) provided with spokes or rays; the hair on the head radiates to form a circular cap, or bonnet. From southern Asia.

Lion-tailed Monkey   *M. silenus*
Silenus was a companion of the Roman god Bacchus and Sileni (plural) were gods of the woods. It has a tuft at the tip of the tail like that of a lion. Inhabiting the forests in the Western Ghats of south-western India.

Pig-tailed Monkey   *M. nemestrina*
Nemestrinus was the god of groves: 'pig-tail' is said to be a reference to the hair on the head, but more likely refers to the tail which is short and 'pig-like'. Ranging from the peninsular part of southern Thailand through Malaya to Borneo, Sumatra and some small adjacent islands.

Crab-eating Macaque   *M. fascicularis*   (formerly *irus*)
fascia (L) (kindred with *fascis*) a band, a filet, hence *fasciculus* (L), the diminutive, a small band   *-aris* (L) suffix meaning pertaining to; Sir Stamford Raffles does not explain the significance of the specific name, but it is probable that he had in mind the coloration of the animal. It inhabits Malaysia, the Philippines, and Indonesia as far east as Flores and Timor. Sir T. Stamford Raffles (1781-1826) was Lieutenant Governer of Java in 1811 and later Sumatra. He founded Singapore in 1819.

Rhesus Monkey   *M. mulatta*
A mulatto is the offspring of a black person mating with a white, from

mulus (L) a mule; this monkey is a yellowish sandy-brown colour *Rhēsos* (Gr) a king of Thrace; it gives the name to the rhesus blood group because experiments in blood transfusion were carried out with this monkey. Inhabiting India, and ranging as far north as the Himalayas, and eastwards to China.

Assamese Macaque   *M. assamensis*
-ensis (L) suffix meaning belonging to, usually applied to localities; belonging to Assam, India.

Formosan or Round-headed Macaque   *M. cyclopis*
kuklos (Gr) round, a circle   *ops*, from *opsis* (Gr) aspect, face; 'round-faced' or 'round-headed'. Inhabiting Formosa.

Stump-tailed Macaque   *M. arctoides*   (formerly *speciosa*)
arktos (Gr) a bear   *-oides* (New L), from *eidos* (Gr) apparent shape, form; 'bear-like'; it is a heavily built very hairy monkey, and has a stumpy tail about 5 cm (2 in) long. Inhabiting south-east Asia.

Japanese Macaque   *M. fuscata*
fuscus (L) dark coloured, or black; it has a dark grey coat. Inhabiting Japan.

Barbary Ape   *M. sylvana*
silva (L) a wood   *-anus* (L) suffix meaning belonging to; Sylvanus was a god of the woods. Although known as apes, they are actually monkeys that have no tail (see page 77). They are the famous monkeys which live on the Rock of Gibraltar and their numbers are maintained at a certain level by the British Government. In their wild state they live in Morocco and Algeria, and it is not known for certain whether they were originally imported or are the last survivors of European monkeys. Barbary is the name for the belt of land north of the Sahara stretching from Egypt to the Atlantic.

Moor Macaque   *M. maurus*
mauros (Gr) dark; it is almost black, and often confused with the Black Ape. Inhabiting the Island of Celebes, Indonesia.

Celebes Black Ape   *Cynopithecus niger*
kuōn (Gr), genitive *kunos*, a dog   *pithēkos* (Gr) an ape   *niger* (L) black the tail is only vestigial but it is not an ape (see page 77). It has a dog like head.

Grey-cheeked Mangabey   *Cercocebus albigena*
kerkos (Gr) the tail   *kēbos* (Gr) a monkey; the mangabeys have lon

tails   *albus* (L) white; can mean pale or whitish   *genus* (Gr) the jaw, cheek. Mangabey is a Malagasy word, derived from the port town of Mangabe; the name is misleading and due to a mistake as these monkeys do not live in Madagascar; they live in Uganda, Congo, Cameroun and other parts of central West Africa.

Black Mangabey   *C. aterrimus*

ater (L) black, hence *aterrimus* (L) very black. Inhabiting forests of southern Congo.

Agile Mangabey   *C. galeritus galeritus*

galeritus (L) wearing a hood or skull cap; it has a distinct crest of hair on the head. Inhabiting forests of central Africa. The nominate subspecies.

Golden-bellied Mangabey   *C. g. chrysogaster*

khrusos (Gr) golden   *gaster* (Gr) the belly. A subspecies of *C. galeritus* (see page 18), it lives in the forests of central and western Africa.

White-collared or Sooty Mangabey   *C. torquatus*

torquatus (L) having a collar; it has a collar of white fur like a ruff, and the coat is usually grey but shows some variation. Inhabiting the forests of central western Africa.

Sacred or Hamadryas Baboon   *Papio hamadryas*

papio (New L) a baboon, from the French *papion*: a hamadryad is a wood nymph from *hama* (Gr) together with, and *drus* (Gr) a tree. Inhabiting hillsides in Arabia, Ethiopia and Sudan.

Yellow Baboon   *P. cynocephalus*

kuōn (Gr), genitive *kunos*, a dog   *kephalē* (Gr) the head; all baboons have a rather dog-like face and muzzle. It inhabits the central part of East Africa.

Chacma Baboon   *P. ursinus*

ursus (L) a bear   *-inus* (L) suffix meaning like   *chacma* is a Hottentot native name. It is found in Rhodesia and South Africa.

Anubis Baboon   *P. anubis*

Supposed to resemble the pictorial representation of Anubis, a god in ancient Egyptian religion. Inhabiting Chad, Sudan, Ethiopia and the northern part of Kenya.

Guinea Baboon   *P. papio*

Inhabiting Guinea and neighbouring areas in western Africa.

Mandrill   *Mandrillus sphinx*
Mandrill is probably a combination of 'man' and 'drill', a West
African native name for a baboon; a man-like ape   the sphinx was
a monster of Egyptian origin figuring also in Greek mythology; there
is probably no special reason for giving it this name (see page 87).
It inhabits tropical rain forests in Gabon and neighbouring areas.

Drill   *M. leucophaeus*
Drill is a West African native name   *leukos* (Gr) white   *phaios* (Gr)
dusky, dark; there is sometimes white hair surrounding the face and
on the chest but most of the coat is dark brown. It lives in the forests
of the Gabon and Cameroun areas in West Africa.

Gelada Baboon   *Theropithecus gelada*
thēr (Gr) a wild animal   *pithēkos* (Gr) an ape   *gelada* is an Ethiopian
native name. It lives in the hills of Ethiopia, and has been seen at
heights of over 2,000 m (7,000 ft) above sea level.

Green or Grass Monkey   *Cercopithecus aethiops*
The monkeys of the genus *Cercopithecus* are known as Guenons; they
have very long tails and colourful fur. The Latin name is misleading
as they are monkeys and not apes (see page 77).
kerkos (Gr) the tail of an animal   *pithēkos* (Gr) an ape   *aithos* (Gr)
burnt, red-brown colour   *opsis* (Gr) aspect, appearance; hence
aethiops = Ethiopian, or negro, i.e. 'burnt-face'. It has a black face
and the coat sometimes has a green tinge. Widespread in Africa south
of the Sahara.

Moustached Monkey   *C. cephus*
cephus is a Linnaean name, a variant of the Greek *kēbos* and similar to
the Arabic *keb*, a monkey, and a name still used in the East African
coastal area. It has a prominent white moustache. Inhabiting central
western Africa.

Diana Monkey   *C. diana*
A white crescent on the forehead is supposed to resemble the bow of
the goddess Diana but this may not be the origin of the name. Inhabiting the central part of western Africa.

Owl-faced Monkey   *C. hamlyni*
R. I. Pocock says that the first specimen of this monkey was 'procured
alive from the Ituri Forest, Congo, for the Hon. Walter Rothschild
by J. D. Hamlyn'. It is possible that Hamlyn was not a collector but

a dealer in animals. The Hon. Walter Rothschild later became Lord Rothschild (see page 132). R. I. Pocock F.R.S. (1863–1947) was Superintendent of the Zoological Gardens in London from 1904 to 1923. The Owl-faced Monkey has large round eyes like those of an owl; it is a rare species living in eastern Congo.

Blue or Diadem Monkey   *C. mitis*
mitis (L) mild, gentle; it is a quiet and shy monkey   *diadema* (L) a royal headband; it has a light coloured band on the forehead. Living in central areas of Africa including Congo, Uganda and Tanzania.

De Brazza Monkey   *C. neglectus*
neglectus (L) neglected, not chosen; so named because the specific rank had been overlooked for some time. J. C. S. de Brazza (1859–1888) was a French naturalist and explorer who spent some time in Congo where this monkey lives. It is also found in parts of Uganda and Kenya.

Talapoin Monkey   *C. talapoin*
Talapoin is the name for a Buddhist monk, particularly of Pegu, in Lower Burma; from *tala poi* (Old Peguan) meaning 'my lord'; it has a dark crown on the head like a monk's cap. Inhabiting some large areas on the west coast of central Africa.

Red-tailed Monkey   *C. ascanius*
Ascanius was the legendary son of Aeneas and Creusa in Greek mythology. This is an olive-green monkey with a red tail living in a small area in the western part of Africa.

Allen's Swamp Monkey   *C. (Allenopithecus) nigroviridis*
The subgenus is named after Dr Joel Asaph Allen (1838–1921) the eminent American zoologist, who was curator of birds and mammals at the American Museum of Natural History, New York, during the years 1885 to 1908   *pithēkos* (Gr) an ape   *niger* (L) black   *viridis* (L) green; it is black and yellowish with a speckled effect. Inhabiting the central part of Africa. The name in brackets indicates that it is a subgenus of *Cercopithecus*.

Patas Monkey   *Erythrocebus patas*
ruthros (Gr) red   *kēbos* (Gr) a monkey; it is a reddish-brown colour *atas*, from *pata* (Wolof); the Wolof are a people of the western Sudan where this monkey lives. It is also found in western parts of Uganda and Kenya.

Hanuman or Entellus Langur   *Presbytis entellus*
presbus (Gr) an old man and so *presbuteros* (Gr) an elder, a priest; can mean greater, more important. The leader, or dominant male of a pack, has a distinct behaviour pattern when issuing orders to his subordinates. Entellus was a massively built veteran hero in Virgil's Aeneid. Hanuman is a Hindu word for a monkey god; this langur is sacred in India. Inhabiting Sri Lanka and peninsular India, ranging north to Sikkim and Kashmir, and to southern Tibet.

Purple-faced Langur   *P. senex*
senex (L) an old man, person; from its appearance, the male having copious whiskers, often white, and the face is reddish to purple. Inhabiting southern India and Sri Lanka.

Spectacled or Dusky Langur   *P. obscurus*
obscurus (L) dark; it has a dark coat and white rings round the eyes, hence 'spectacled'. Inhabiting southern Burma, southern Thailand, Malaga and some small adjacent islands.

Maroon Langur   *P. rubicundus*
rubicundus (L) red, ruddy. Inhabiting Borneo.

White-headed Langur   *P. leucocephalus*
leukos (Gr) white   *kephalē* (Gr) the head; it is possible that this langur will be established as a subspecies; it lives in a small area in Kwangsi, to the north of Tongking.

Douc Langur   *Pygathrix nemaeus*
pugē (Gr) the rump, buttocks   *thrix* (Gr) hair; alluding to the long hair on the rump   *nemus* (L) a grove, a forest   *douc* is a French Cochin-china name. Cochin-china was a name used by the French in the 17th century for the central and southern areas of Vietnam and the capital was Saigon. This langur inhabits southern Vietnam and Laos, and is now very rare.

Golden Snub-nosed Monkey   *Rhinopithecus roxellanae*
rhis (Gr), genitive *rhinos*, the nose   *pithēkos* (Gr) an ape. This snub nosed monkey is named after the consort of the great Turkish sultan Sulaiman I; she was a famous Russian lady of doubtful repute, named Roxellane; like the monkey she had golden-red hair and a turned-up nose. This is a very rare monkey discovered in 1870, and living in th mountains of eastern Tibet and north-eastern China.

Pig-tailed Langur   *Simias concolor*
simias (L) an ape, a monkey   *concolor* (L) the same colour. This langur's tail is short and naked; the coat is entirely brown but the face,
hands and feet are black. Inhabiting the mountains of the Mentawei
Islands, off the west coast of Sumatra. It is very rare.

Proboscis Monkey   *Nasalis larvatus*
nasus (L) the nose and so *nasalis* (New L) pertaining to the nose   *larva*
(L) a ghost; can mean a mask   *-atus* (L) suffix meaning provided
with, 'having a mask'; this is a reference to the nose, which is long and
quite remarkable. It lives in the swamp forests of Borneo.

Black Colobus   *Colobus polykomos*
kolobos (Gr) docked, mutilated; they have very small thumbs or the
thumb completely absent, giving the appearance that it has been cut
off   *polus* (Gr) many   *komē* (Gr) the hair; the coat is handsomely
marked in black and white and long and silky. Widespread in the
tropical forests of Africa.

Red Colobus   *C. badius*
badius (L) chestnut coloured; it has a black body and chestnut
coloured head, arms and legs. Range as above.

Olive Colobus   *C. verus*
verus (L) true, genuine; meaning a true or typical colobus. It is confined to an area in and around Ghana on the west coast of Africa.

Family   HYLOBATIDAE   7 species
hulē (Gr) a wood, a forest   *bainō* (Gr) I walk, I step   *batēs* (Gr) one
that treads; can mean a climber.

White-handed or Lar Gibbon   *Hylobates lar*
lar is an honorary title in Rome, equivalent to the English Lord.
The Lar Gibbon lives in Thailand, the Malay Peninsula, Sumatra,
Java and Borneo.

Dark-handed Gibbon   *H. agilis*
agilis (L) nimble, agile; the gibbons are probably the most expert
acrobats in the animal kingdom. This gibbon has a dark upper surface
on the hands and feet; it inhabits the Malay Peninsula and Sumatra.

Hoolock Gibbon   *H. hoolock*
hulluk is a Burmese native name for the gibbon; it inhabits Burma and
Thailand.

Black or Harlan's Gibbon   *H. concolor concolor*
concolor (L) the same colour   Dr Richard Harlan (1796-1843) was
an American physician, naturalist and author. This gibbon inhabits
Cambodia, Vietnam and Laos; the nominate subspecies.

White-cheeked Gibbon   *H. c. leucogenys*
leukos (Gr) white   *genus* (Gr) the jaw, cheek; the cheeks are white or
pale yellow. This is a subspecies of *H. concolor* living in Vietnam, Laos
and Thailand.

Grey Gibbon   *H. moloch*
Moloch was a Semitic god to whom children were sacrificed; the
specific name *moloch*, given by Audebert in 1797, probably has no
special significance. The custom of naming monkeys and apes after
mythical creatures and classical and heathen beings probably derives
from Linnaeus (see page 87). Jean Baptiste Audebert (1759-1800)
was a distinguished naturalist, bird artist and engraver. This gibbon
inhabits Java and Borneo.

Siamang   *Symphalangus syndactylus*
sym- (= *syn-*) (Gr) together   *phalanx* (Gr), genitive *phalangos*, soldiers
in line of battle; in a biological sense the small bones between the
joints of the fingers, and also the toes; in the siamang the second and
third toes are joined together by a web of skin   *daktulos* (Gr) a finger
also a toe; so this name means 'the toes-joined one, with joined toes'
Inhabiting Sumatra.

Dwarf Siamang   *S. klossi*
C. B. Kloss (1877-1949) was a zoologist living in Singapore during
the years 1903 to 1932   *siamang* is a Malay native name for the gibbon
Some authorities classify this small gibbon in the genus *Hylobates*
It was discovered in the early part of the twentieth century and i
found only in the Mentawai Islands off the west coast of Sumatra

Family   PONGIDAE   4 species
mpongi is a Congolese name, probably originally used to mean
gorilla, but later used to mean the orang-utan; this is a Malay word
meaning 'forest-man'.

Orang-Utan   *Pongo pygmaeus*   (*Pongo* formerly *Simia*)
pugmaios (Gr) small, dwarfish; a misleading name as it is quite a larg
ape, standing about four feet high; the comparison was probabl
made with a man. Inhabiting Borneo and Sumatra.

Chimpanzee *Pan troglodytes* (*Pan* formerly *Anthropopithecus*)
pan (Gr) all, the whole; in Greek mythology Pan was the rural god
of Arcadia, of pastures and woods *trōglē* (Gr) a hole *dutēs* (Gr) a
burrower, a diver; a peculiar name for the chimpanzee as a troglodyte
is a cave-dweller, whereas chimpanzees spend some of their time in
trees and make nests there where they sleep, and some of their time
on the ground; they do not live in caves. The name was given nearly
200 years ago by J. F. Blumenbach, in 1779, and in those days it was
probably thought to be some kind of cave man. It inhabits widely
scattered areas of central parts of western Africa north of the Zaire
River (formerly Congo River), particularly where there are tall
deciduous forests.

Pygmy Chimpanzee *P. paniscus*
-*iscus* (L) dim, suffix, 'a small Pan'; it is also known as the Dwarf
Chimpanzee, and is slender and much smaller than *P. troglodytes*. The
name chimpanzee is from a Zaire native name *kimpenzi*. It lives in the
swamp forests of Zaire, south of the Zaire River and ranging southwards to the Lukenie River.

Western Gorilla *Gorilla gorilla gorilla*
gorillai (Gr) gorillas or hairy humans; the origin of the word is obscure.
Hanno the Carthaginian said it was a tribe of hairy women; the name
almost certainly is from western Africa. This gorilla inhabits the
forests of Cameroun and Gabon in central western Africa. (See
Tautonyms on page 13.) It is the nominate subspecies.

Mountain Gorilla *G. g. beringei*
Discovered by Capt Oskar von Beringe, an officer in what was then
German East Africa, in 1902. It was named by Dr Paul Matschie
(1862-1926) the German zoologist, and originally misspelled as
eringeri. It lives high up in the mountains to the west of Lake Kivu
in eastern Zaire, and contiguous regions of Uganda, Rwanda,
Burundi and Tanzania. A subspecies.

Family HOMINIDAE 1 species
Homo (L), genitive *hominis*, a man.

Man *Homo sapiens*
sapiens (L) wise, sensible(!). Widespread throughout the world.

THE PRIMATES

Their relationship to humans and other animals, showing the Phylum, Subphylum, C
Subclass, Order, Suborders, and Families, in sequence.

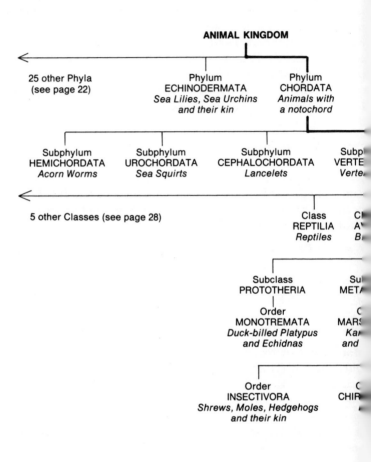

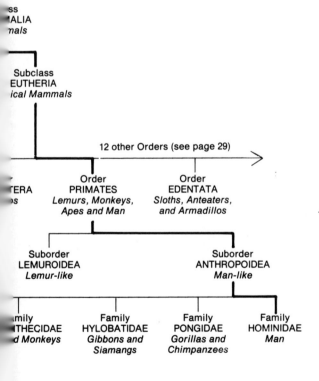

ss
ALIA
mals

Subclass
EUTHERIA
ical Mammals

12 other Orders (see page 29)

ERA
s

Order	Order
PRIMATES	**EDENTATA**
Lemurs, Monkeys,	*Sloths, Anteaters,*
Apes and Man	*and Armadillos*

Suborder
LEMUROIDEA
Lemur-like

Suborder
ANTHROPOIDEA
Man-like

mily
THECIDAE
d Monkeys

Family
HYLOBATIDAE
*Gibbons and
Siamangs*

Family
PONGIDAE
*Gorillas and
Chimpanzees*

Family
HOMINIDAE
Man

12 Sloths, Armadillos and Anteaters
EDENTATA

This is a curious little group of mammals coming under the order Edentata, and even this name makes no sense as it means 'without teeth', whereas in fact most of them do have teeth, and the Giant Armadillo has up to ninety! The name came about because it was first applied to the toothless anteaters. These have no teeth and feed exclusively on termites and other insects. (They should not really be known as 'anteaters' because, strictly speaking, termites are not ants. They are distantly related to ants, but are not classed as the same order, termites being in the order Isoptera, meaning 'equal or similar wing', and ants in the order Hymenoptera, meaning 'membrane wing'.) With regard to the sloths, there are only two known types, although these are divided into several species; the nine neck vertebrae of the Three-toed Sloth is unique, being more neck vertebrae than even the giraffe.

Subclass EUTHERIA
Order EDENTATA
e (=*ex*) (L) prefix meaning out of; can mean without   *dens* (L), genitive *dentis*, a tooth   *-atus* (L) suffix meaning provided with; 'not provided with teeth'.

Family MYRMECOPHAGIDAE   4 species
murmex (Gr) genitive *murmēkos*, an ant   *phagein* (Gr) to eat.

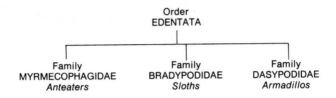

Giant Anteater   *Myrmecophaga tridactyla*
treis or *tria* (Gr) three   *daktulos* (Gr) a finger, a toe. Inhabiting
Central and South America.

Silky or Two-toed Anteater   *Cyclopes didactylus*
kyklops (Gr) round-eyed, or possibly *kuklos* (Gr) round, a circle   *pes*
(L) a foot; may refer to this anteater's unusual feet which have a
jointed sole that enables the claws to bend round and almost completely encircle the branch of a tree   *dis* or *di-* (Gr) twice, double
daktulos (Gr) a finger, a toe. Inhabiting Central America and the
northern part of South America.

Tamandua   *Tamandua tetradactyla*
tamandua is a Brazilian word for an 'ant-trap'   *tetras* (Gr) four
daktulos (Gr) a finger, a toe. Inhabiting Central and South America.

Family   BRADYPODIDAE   probably 7 species
bradus (Gr) slow   *pous* (Gr), genitive *podos*, a foot; hanging from the
branches of trees, they usually move very slowly, hence the name
sloth.

Three-toed Sloth or Ai   *Bradypus tridactylus*
treis, tria (Gr) three   *daktulos* (Gr) a finger, a toe. It has three toes on
the forefeet and hind feet. It inhabits Central America and most of
the northern half of South America.

Two-toed Sloth or Unau   *Choloepus didactylus*
khōlos (Gr) lame, maimed   *pous* (Gr) a foot; this name has been given
because it has only two digits on the forefeet; the name is misleading
because it has three toes on the hind feet   *dis* or *di-* (Gr) twice, double
daktulos (Gr) a finger, a toe. Unau and Ai (above) are Tupi native

names for the sloth. It inhabits Central America and the north-western part of South America.

Hoffmann's Sloth *C. hoffmanni*
Dr Carl Hoffmann (1823–1859) was in Costa Rica, Central America, from 1854 until his death in 1859. This two-toed sloth inhabits Costa Rica and ranges south to Ecuador in South America.

Family DASYPODIDAE 21 species
dasus (Gr) hairy, rough *pous* (Gr), *genitive* podos, a foot; in this case meaning 'rough-footed'.

Long-haired Armadillo *Chaetophractus vellerosus*
khaitē (Gr) long flowing hair *phraktos* (Gr) protected vellus (L), genitive *velleris*, wool, hair *-osus* (L) suffix meaning full of. Inhabiting South America.

Giant Armadillo *Priodontes giganteus*
priōn (Gr) a saw *odous* (Gr), genitive *odontos*, a tooth; 'saw-toothed'; they have up to ninety small teeth: giganteus (L) very big. Inhabiting forested areas of the Amazon basin.

Central American Armadillo *Cabassous centralis*
Cabassous is probably from *capacou* (Galibi) an armadillo; the native language of a people of French Guiana *-alis* (L) suffix meaning relating to. From Central America.

La Plata Three-banded Armadillo *Tolypeutes matacus*
tolupē (Gr) wool made into a ball, a ball wound up *-tes* (Gr) suffix meaning in connection with; these small armadillos are the only ones that roll themselves up to form a ball when in danger; the Mataco are a people of Bolivia, Paraguay and Argentina *-acus* (L) suffix meaning relating to. It lives in Paraguay and parts of Bolivia and Argentina.

Brazilian Three-banded Armadillo *T. tricinctus*
tria (L) three *cinctus* (L) a girdle; there are three belts of bony scales round the body. Inhabiting the eastern part of Brazil.

Nine-banded Armadillo *Dasypus novemcinctus*
dasus (Gr) hairy, rough *pous* (Gr) a foot; in this case it means 'rough-footed' *novem* (L) nine *cinctus* (L) a girdle; it has nine conspicuous belts of bony scales round the body. It inhabits the southern part of North America and is widespread in South America.

Fairy Armadillo or Pichiciego   *Chlamyphorus truncatus*
khlamus (Gr) a short cloak or mantle   *phora* (Gr) a carrying; can also
mean that which is carried, a burden; referring to the armour plating
of bony scales on the armadillo   *trunco* (L) I maim, I cut off, hence
truncatus (New L) cut off; the armour plating on the back stops
abruptly above the rump and the rear end of the body is protected
by a bony shield. It is the smallest of the twenty one species, being
only about 21 cm (6 in) long; it inhabits Argentina and Bolivia.
Pichiciego is the local Allentiac name in Western Argentina.

13 **Pangolins** PHOLIDOTA

The Pangolins are in some ways similar to the armadillos and ant-eaters, though not actually related to them. The head, back and upper part of the tail are covered in overlapping epidermal scales; they have an amazingly long tongue for insertion into anthills, in some cases nearly as long as the body, and a long prehensile tail. The name is from a Malayan word *peng-goling*, a roller, on account of the habit the animal has of rolling itself up to form a ball when in danger.

Subclass EUTHERIA
(see pages 32 and 51)

Order PHOLIDOTA
pholis (Gr) genitive *pholidos*, a horny scale; the head, back, and upper part of the tail are covered with horny scales.

Family MANIDAE 7 species
Derived from *manes* (L) in Roman religion the spirits of the dead, or ghosts; so named because of their nocturnal habits and peculiar appearance.

Giant Pangolin *Manis gigantea*
Manis is an assumed Latin singular, coined from *manes* (see above) *gigas* (L) a giant, hence *giganteus* (L) big, gigantic. It inhabits the western part of Africa south of the Sahara.

Order
PHOLIDOTA
|
Family
MANIDAE
Pangolins

Short-tailed or Cape Pangolin *M. temmincki*
Professor C. J. Temminck (1778–1858) was a Dutch zoologist, and at one time Director of the Natural History Museum at Leyden in the Netherlands. This pangolin lives in South Africa.

Small-scaled Tree Pangolin *M. tricuspis*
tres, *tria* (L) three *cuspis* (L) a point; the edges of the scales on the young pangolin have three points, though these disappear in the adult. Inhabiting the western part of Africa, south of the Sahara, and ranging eastwards into Uganda.

Long-tailed Pangolin *M. tetradactyla* (*tetradactyla* or *longicauda*)
tetra (Gr) four *daktulos* (Gr) a finger, toe. The tail is about twice the length of the head and body which is most unusual in mammals. It is sometimes known as the Black-bellied Pangolin from the dark hair on the underparts. Range as for *M. tricuspis*.

Chinese Pangolin *M. pentadactyla*
pente (Gr) five *daktulos* (Gr) a finger, toe. Inhabiting Nepal, China, Hainan and Formosa.

Indian Pangolin *M. crassicaudata*
crassus (L) thick, heavy *cauda* (L) the tail *-atus* (L) suffix meaning provided with; it has a very broad heavy tail. It lives in southern Asia

Malayan Pangolin *M. javanica*
Inhabiting Burma, Java, Sumatra, Borneo, Celebes and southern parts of the Philippines.

It should be noted that three or four subgenera are often recognised for example *Phataginus* and *Smutsia*; phatagen is an East Indian name for the pangolin, and Johannes Smuts was an early nineteenth century South African naturalist.

Until fairly recently the hares and rabbits were classified with the rodents, partly on account of the teeth, and also perhaps the shape of the head. However, they have four upper incisors instead of two, as in the rodents, and the upper and lower jaws oppose each other only on one side at a time, thus the motion of chewing is lateral instead of longitudinal. They are now placed in a separate order, Lagomorpha, and this includes the furry little animals known as pikas, sometimes called piping-hares on account of the shrill whistling and calling noises they make.

Domestic rabbits derive from the wild species *Oryctolagus cuniculus*, and by selective breeding in captivity many varieties of this one species have been produced.

Subclass **EUTHERIA**
(see pages 32 and 51)

Order **LAGOMORPHA**
lagōs (Gr) a hare *morphē* (Gr) the form, shape.

Family **OCHOTONIDAE** 14 species
Ochotona (New L) derived from *ochodona*, a Mongolian name for the pika.

Russian or Steppe Pika *Ochotona pusilla*
pusillus (L) very small. The name pika originates from *piika* (Tun-

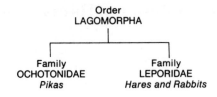

gusic). The Tungus are a people of eastern Siberia. Inhabiting Russia and Asia.

Large-eared Pika   *O. macrotis macrotis*
makros (Gr) long   *ous* (Gr), genitive *ōtos*, the ear. From Asia.

Mount Everest Pika   *O. m. wollastoni*
Named after Dr A. F. R. Wollaston (1875–1930), an English naturalist and explorer; it was found at an altitude of 5,300 m (17,500 ft). A subspecies of *O. macrotis* above.

Pallas's Pika   *O. pallasi*
Named after Professor Peter Simon Pallas (1741–1811), a German zoologist and explorer and a professor at St Petersburg University in Russia. He made contributions to most of the natural sciences. This pika inhabits the Volga district and the Ural Mountains in Russia.

American or Rocky Mountain Pika   *O. princeps*
princeps (L) a chief; a reference to an Amerindian name translated as 'Little Chief Hare'. Inhabiting North America.

Family   LEPORIDAE   about 50 species
lepus (L), genitive *leporis*, a hare.

Ryukyu Rabbit   *Pentalagus furnessi*
pente (Gr) five (in composition *penta-*)   *lagōs* (Gr) a hare; an allusion to the five pairs of upper cheek teeth instead of the usual six: 'Collected on the Liu Kiu Islands by Dr W. H. Furness and Dr H. M. Hiller on Feb 26th 1896.' No further information about these two collectors seems to be available. Inhabiting the Ryukyu Islands that lie to the south of Japan in the Pacific Ocean, it is jealously preserved as a 'Natural Monument'.

Natal Red Hare   *Pronolagus crassicaudatus*
pronus (L) leaning forward; can mean belonging to what is before;
this hare has characters of an earlier form   *crassus* (L) thick, heavy
cauda (L) the tail of an animal. Inhabiting South Africa.

Volcano Rabbit   *Romerolagus diazi*
Romero is a town in Mexico that lies in the volcanic belt that crosses
the country from east to west, and consisting of famous volcanos such
as Popocatapetl. However, the name is in honour of Don Matias
Romero (1837–1898) in recognition of his assistance to the Biological
Survey in Mexico; he was a Government Minister. The name *diazi*
commemorates D. A. Diaz de Leon and was given by Dr Jesus Diaz
de Leon. This rabbit is found only on the slopes of two volcanos lying
to the south-east of Mexico City.

Assam Rabbit   *Caprolagus hispidus*
kapros (Gr) a wild boar   *lagōs* (Gr) a hare; probably an allusion to
the coarse, bristly fur   *hispidus* (L) rough, hairy. This rabbit is confined to a small area in Assam, and at one time was thought to be
extinct, but a few have been seen recently and it is now to be protected.

Cape Hare   *Lepus capensis capensis*
-ensis (L) suffix meaning belonging to; it is not confined to the Cape
area, South Africa, and ranges through Europe and parts of Asia.

European or Brown Hare   *L. c. europaeus*
The hare of the British Isles and Europe, formerly *Lepus europaeus*, is
now considered to belong to the Afro-Mediterranean species complex
of *L. capensis* (above). Thus, by the law of priority *L. capensis* named
by Linnaeus in 1758, must take precedence over *L. europaeus* named
by Pallas in 1778, 20 years later. This means it is now a subspecies,
Lepus capensis europaeus. It ranges through Europe, Africa and western
Asia.

Alpine Hare   *L. timidus timidus*
timidus (L) afraid, timid. Living in the Alps of Europe and the
Scandinavian mountains, and ranging eastwards as far as Japan.

Scottish or Blue Hare   *L. t. scoticus*
icus (L) suffix meaning belonging to. This subspecies of *L. timidus*
lives in the mountains of Scotland and ranges south to northern
England and Wales; it is sometimes known as the Varying Hare
because the coat becomes white in winter.

Snowshoe Rabbit   *L. americanus*
-*anus* (L) suffix meaning belonging to. It is not a rabbit, but a hare;
stiff bristles grow on the feet in autumn to help when running on snow
and ice, hence the name 'snowshoe'. It inhabits North America.

White-tailed Jack Rabbit   *L. townsendi*
This is a hare with the typical long hind legs. The names 'hare' and
'rabbit' become rather indiscriminately mixed, so that some animals
known as rabbits are, in fact, hares. J. K. Townsend (1809-1851) was
an ornithologist and author who was exploring in the Rockies in 1834.
It inhabits the Rocky Mountains area in the north-west part of North
America.

Black-tailed Jack Rabbit   *L. californicus*
-*icus* (L) suffix meaning belonging to. Like some other hares, known
as a 'varying hare' because its coat becomes white in winter. This
'Jack Rabbit' is a hare inhabiting North America.

Cottontail   *Sylvilagus floridanus*
silva (L) a wood   *lagōs* (Gr) a hare; in spite of this name it is a rabbit
-*anus* (L) suffix meaning belonging to; of Florida. The tail is white
underneath, and looks like a ball of white cotton when raised; this is
used as an alarm signal and also in certain courting ceremonies. It is
not confined to Florida, and ranges through Canada, the USA,
Central America and the northern part of South America.

Pygmy Rabbit   *S. idahoensis*
-*ensis* (L) suffix meaning belonging to. This is the smallest rabbit, only
about 25 cm (10 in) long; it is found in Idaho and other parts of North
America.

Marsh Rabbit   *S. palustris*
paluster (L), genitive *palustris*, marshy, boggy. Inhabiting the south
eastern part of North America.

Swamp Rabbit   *S. aquaticus*
aqua (L) water   -*icus* (L) suffix meaning belonging to, hence *aquaticu*
(L) living in water; it likes swampy conditions. Inhabiting the
southern part of North America.

European Rabbit   *Oryctolagus cuniculus*
oruktēr (Gr) a tool for digging   *lagōs* (Gr) a hare; 'a digging hare'
i.e. a rabbit, as hares do not make burrows   *cuniculus* (L) a rabbit
can also mean an underground passage. This is the common rabbi

known throughout Britain and Europe. It is essentially a burrowing animal. Originally inhabiting Europe including the British Isles, and North Africa, it has now been introduced to many other countries.

Short-eared Rabbit   *Nesolagus netscheri*

nēsos (Gr) an island   *lagōs* (Gr) a hare; 'an island hare': named after E. Netscher, a naturalist who was at one time a member of the Council of Dutch East Indies (now Indonesia). This is a hare, but with remarkably short hind legs and short ears; it is sometimes known as the Sumatran Hare. A rare animal, found only in the tropical forest areas on the Island of Sumatra, Indonesia.

15 Rodents RODENTIA

This is a large group, consisting of gnawing animals and containing over 1,500 species. It includes the mice, rats, guinea pigs, hamsters, squirrels, beavers, porcupines and many other less well-known rodents. They are, on the whole, vegetarians, and their teeth are specially adapted for such a diet. However, some will eat insects; the tooth formation is an infallible guide in identifying them as rodents. They have a single pair of continuously growing incisors in both upper and lower jaws, but have no canines.

For purposes of classification the order Rodentia has been divided into three suborders: Sciuromorpha, the 'squirrel-like'; Myomorpha, the 'mouse-like'; Hystricomorpha, the 'porcupine-like'.

Subclass EUTHERIA
Order RODENTIA
rodo (L) I gnaw.

Suborder SCIUROMORPHA
sciurus (L) a squirrel; also *skiouros* (Gr) a squirrel. This is derived from *skia* (Gr) shade, and *oura* (Gr) the tail; a 'shade-tail', on account of the way a squirrel holds his bushy tail over his back *morphē* (Gr) form, shape, resemblance.

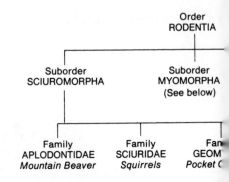

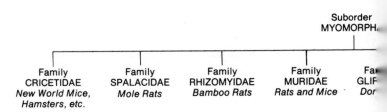

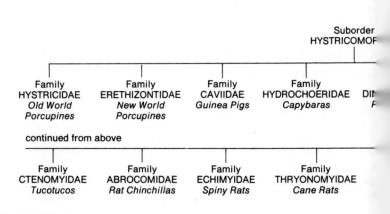

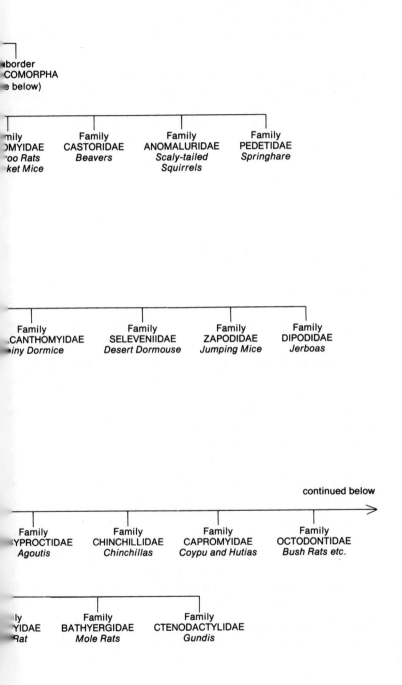

Suborder
MYOMORPHA
(continued below)

Family
HETEROMYIDAE
Kangaroo Rats
Pocket Mice

Family
CASTORIDAE
Beavers

Family
ANOMALURIDAE
Scaly-tailed
Squirrels

Family
PEDETIDAE
Springhare

Family
PLATACANTHOMYIDAE
Spiny Dormice

Family
SELEVENIIDAE
Desert Dormouse

Family
ZAPODIDAE
Jumping Mice

Family
DIPODIDAE
Jerboas

continued below

Family
DASYPROCTIDAE
Agoutis

Family
CHINCHILLIDAE
Chinchillas

Family
CAPROMYIDAE
Coypu and Hutias

Family
OCTODONTIDAE
Bush Rats etc.

Family
CTENOMYIDAE
Tuco Tuco Rat

Family
BATHYERGIDAE
Mole Rats

Family
CTENODACTYLIDAE
Gundis

Family   APLODONTIDAE   1 species
aploos (Gr) simple, single   *odous* (Gr), genitive *odontos*, a tooth; the
cheek teeth are simple; they have no roots and the molar teeth have
single crowns.

Mountain Beaver or Sewellel   *Aplodontia rufa*
rufus (L) red, ruddy. In spite of the name, it is not a beaver, though
it is fond of water and swims well. The name sewellel comes from a
Chinook word *shewallal*, which means a cloak made from the animal's
skin. There is only one species, and it inhabits the western side of
North America.

Family   SCIURIDAE   250 species, probably more
sciurus (L) see above.

Red Squirrel   *Sciurus vulgaris*
sciurus (L), see above   *vulgaris* (L) common, ordinary. Widespread
in the forests of Europe and Asia.

Grey Squirrel   *S. carolinensis*
-ensis (L) belonging to; it is not confined to Carolina, from where it
takes its name, and is widespread in North America, and latterly also
in Great Britain.

American Red Squirrel or Chickaree   *Tamiasciurus hudsonicus*
tamias (Gr) a treasurer, one who stores; 'a hoarder'   *sciurus*, see above
-icus (L) suffix meaning belonging to; it is not confined to the Hudson
Bay area, and is widespread in North America. It should not be
confused with the European Red Squirrel, *Sciurus vulgaris*. Chickaree
is said to be an imitation of its cry.

Douglas's Squirrel   *T. douglasii*
Named after David Douglas (1798-1834) a botanist and explorer of
North America, it lives in the California area.

Indian Palm Squirrel   *Funambulus palmarum*
funis (L) a rope and *ambulo* (L) I walk, hence funambulus (L) a rope
dancer, a tight-rope performer   *palma* (L) the palm tree, hence
palmarius (L) of palms; often seen in palm forests because they are
fond of oil-palm nuts, they have other quite varied habitats. Widespread in southern Asia.

Indian Giant Squirrel   *Ratufa indica*
Ratuphar is a native name for this squirrel in Monghyr, a district

Bengal   -icus (L) suffix meaning belonging to. It is a real giant among squirrels with an overall length of about 1 m (3 ft). Inhabiting southern Asia, Sumatra, Java and Borneo.

African Giant or Oil-palm Squirrel   *Protoxerus stangeri*
prōtos (Gr) first, primary; could indicate the first one to be discovered *xerōs* (Gr) dry, parched; so called from the character of the fur which is harsh and often spiny; the name can be misleading as it does not have the bristly hair of *Xerus erythropus*. Dr W. Stanger (1812-1854) was an English scientist and explorer. This squirrel inhabits the western part of Africa.

African Palm Squirrel   *Epixerus ebii*
epi- (Gr) upon, near; indicating it is anatomically like *Protoxerus* (above)   Ebo, a people of southern Nigeria. Inhabiting the western part of Africa.

Sun Squirrel   *Heliosciurus gambianus*
hēlios (Gr) the sun   *skiouros* (Gr) a squirrel; so called from its tropical habitat   -anus (L) suffix meaning belonging to; 'of Gambia'. A brightly coloured Sun Squirrel inhabiting Africa.

African Pygmy Squirrel   *Myosciurus pumilio*
mus (Gr) a mouse, genitive *muos*   *skiouros* (Gr) a squirrel   *pumilio* (L) a dwarf; it is only about 20 cm (4 in) long overall (c.f. *Ratufa indica*, above). Inhabiting the western part of Africa.

Hog or Long-snouted Squirrel   *Hyosciurus heinrichi*
hus (Gr), genitive *huos*, a pig   *skiouros* (Gr) a squirrel; a reference to the long nose   Gerd Heinrich (born 1896) was a collector in Celebes from 1930 to 1932. This squirrel inhabits Celebes.

Striped Ground Squirrel   *Xerus erythropus*
xerōs (Gr) dry, parched; a reference to the fur which is harsh and often spiny   *eruthros* (Gr) red   *pous* (Gr) a foot. Inhabiting savanna country of Africa from Mauritania to Uganda and Kenya.

South African Ground Squirrel   *Xerus (Geosciurus) inauris*
gē (Gr) earth, ground   *skiouros* (Gr) a squirrel   *in-* (L) prefix meaning not, without   *auris* (L) the ear; it has ears, but they are very small and usually hidden under the fur. Inhabiting South Africa. This is a subgenus of *X. erythropus* above.

Alpine Marmot   *Marmota marmota*   (*Marmota* formerly *Arctomys*)
marmotta (It) derived from *murmont* (Romansch, an Upper Rhine

dialect), which was derived from *mus* (L), genitive *muris*, a mouse; and *mons* (L), genitive *montis*, a mountain; 'mountain mouse'; it lives in the mountains, sometimes at an altitude of 2,500 m (8,000 ft) in central and north-eastern Europe.

Bobak   *M. bobak*
bobak (Pol) a marmot. Inhabiting the Himalayas and other mountains in that part of Asia.

Hoary Marmot   *M. caligata*
caliga (L) a boot, hence *caligatus* (L) wearing boots; the lower legs and feet are black, giving the appearance of boots. 'Hoary' indicates a greyish-white coat. Living in the mountains of Alaska and the western part of North America.

Woodchuck or Marmot   *M. monax*
Monax is an American Indian name for the marmot; it lives high up in the mountains of Canada and the USA.

Black-tailed Prairie Marmot   *Cynomys ludovicianus*
kuōn (Gr), genitive *kunos*, a dog   *mus* (Gr) a mouse; sometimes known as the Prairie Dog on account of its sharp barking alarm call, it is not a dog: *ludovicianus* is an adjective connected with the name Louis; it indicates that the animal was found in Louisiana, a southern state in the USA, and it is the area where this marmot lives.

Antelope Ground Squirrel   *Ammospermophilus harrisi*
ammos (Gr) sand, also a sandy place   *sperma* (Gr) seed   *philos* (Gr) loved, pleasing; 'seed-loving one of sandy places'; it lives in the hot sandy desert, and the diet is mostly seed and other plant life. It was named in honour of Edward Harris (1799–1863) who accompanied J. J. L. Audubon (1785–1851), the famous American ornithologist, on his Missouri River trip in 1843. The name, given by Audubon, commemorates their friendship, and several birds also bear his name, for example Harris's Woodpecker *Dryobates villosus harrisi*. The name 'antelope' derives from the tail which shows a white patch when raised, as with the Pronghorn Antelope. This squirrel inhabits desert areas of the USA in the south-west.

White-tailed Antelope Ground Squirrel   *A. leucurus*
leukos (Gr) white   *oura* (Gr) the tail. Inhabiting northern Mexico and Arizona, USA.

Rock Ground Squirrel   *Otospermophilus beecheyi*

ous (Gr), genitive *ōtos*, the ear; the ears are more prominent than in allied genera   *sperma* (Gr) seed   *philos* (Gr) loved, pleasing; the diet is mostly seed and other plant life: named after Rear Admiral F. W. Beechey (1796–1856) who was at one time President of the Royal Geographical Society. This squirrel lives in the California area.

European Ground Squirrel or Souslik   *Citellus citellus*

(Citellus = Spermophilus)

citellus (L) a ground squirrel   suslik is a Russian name for the Ground Squirrel. Inhabiting Europe and Asia.

Thirteen-striped Ground Squirrel   *C. tridecemlineatus*

tria (L) three   *decem* (L) ten   *lineatus* (L) lined. Widespread in the central area of North America.

Barrow Ground Squirrel   *C. parryi*

Dr C. C. Parry (1823–1890) was an American botanist and explorer who is commemorated in several plant names, for example the lily *Lilium parryi*. This ground squirrel lives in Alaska, and takes its English name from Barrow Point, on the northern coast of Alaska; this was named by Sir John Barrow (1764–1848) the English Arctic explorer.

Eastern Chipmunk   *Tamias striatus*

tamias (Gr) a treasurer, one who stores; 'a hoarder'   *stria* (L) a furrow, hence *striatus*, striped; it has stripes along the back. It inhabits the eastern part of Canada and the USA.

Siberian Chipmunk   *Eutamias sibiricus*

eu- (Gr) a prefix meaning well, nicely; usually used to indicate typical *tamias*, see above   *-icus* (L) belonging to; of Siberia.

European Flying Squirrel or Polatouche   *Sciuropterus russicus*

sciurus (L) a squirrel   *pteron* (Gr) a wing   *-icus* (L) belonging to, hence *russicus*, of Russia. The flying squirrels cannot fly in the true sense of the word, but are able to glide a distance of 30 m (100 ft) or more by means of a membrane of skin stretched between the forelegs and the ankles. This squirrel ranges all the way from Scandinavia to Japan. *Polatouche* (Fr) is derived from the Russian *poletusha*, a flying squirrel.

Red and White Flying Squirrel   *Petaurista alborufus*

petauron (Gr) a perch, a springboard   *-ista* (L) suffix denoting ability, one who practises; 'a springboard jumper'   *albus* (L) white   *rufus* (L) red; it has reddish-brown rings round the eyes on a white face. Inhabiting eastern Asia.

American Flying Squirrel   *Glaucomys volans*
glaukos (Gr) silvery, gleaming; can mean grey   *mus* (Gr) a mouse: *volans* (L) flying; like all the flying squirrels, it cannot fly in the true sense of the word, but is able to glide quite long distances by means of a membrane of skin stretched between the forelegs and the ankles. Inhabiting Central and North America.

Family   GEOMYIDAE   about 38 species
gē (Gr) earth, ground   *mus* (Gr) a mouse; referring to the animal's subterranean mode of life.

Pocket Gopher   *Geomys bursarius*
bursa (Gr) the skin stripped off, a hide; giving rise to *bursa* (New L) a pouch, a pocket made of skin   *-arius* (L) suffix meaning belonging to; it has fur-lined pockets on the outside of the cheeks for carrying food. Inhabiting the western part of Canada, the USA and Central America.

Northern Pocket Gopher   *Thomomys talpoides*
thōmos (Gr) a heap   *mus* (Gr) a mouse; an allusion to the heaps of earth thrown out at frequent intervals along the line of the burrows *talpa* (L) a mole   *-oides* (New L) from *eidos* (Gr) apparent shape, resemblance; 'mole-like' because of its burrowing habits. Found in south-western Canada and north-western USA. The genus *Thomomys* has many subspecies.

Family   HETEROMYIDAE   70 species, possibly more
heteros (Gr) the other; can mean different from the usual   *mus* (Gr) a mouse; it is different from *Mus*.

Pocket Mouse   *Perognathus penicillatus*
pēra (Gr) a pouch, a pocket   *gnathos* (Gr) the jaw, the mouth; it has fur-lined pockets on the outside of the cheeks where it can store and carry food   *penicillus* (L) a painter's brush   *-atus* (L) suffix meaning provided with; the tail has a tuft at the end like a painter's brush. It inhabits the western part of Canada, the USA and Central America

Kangaroo Mouse   *Microdipodops megacephalus*
mikros (Gr) small   *di-* from *dis-* (Gr) two, double   *pous* (Gr), genitive

podos, a foot   *opsis* (Gr) appearance; 'one that appears to have two small feet'; the front legs are very small and not used for moving about *megas* (Gr) big   *kephalē* (Gr) the head. Inhabiting North and Central America.

Giant Kangaroo Rat   *Dipodomys ingens*
di- from *dis-* (Gr) two   *pous* (Gr), genitive *podos*, a foot   *mus* (Gr) a mouse; *ingens* (L) vast, enormous; that is compared with other species. Inhabiting Mexico and ranging north to California.

Merriam's Kangaroo Rat   *D. merriami*
Dr C. Hart Merriam (1855-1942) was a zoologist who studied mammals and birds of America, and was Chief of the US Biological Survey in 1885. This kangaroo rat inhabits western and southern parts of North America.

Black-tailed Spiny Pocket Mouse   *Heteromys nigricaudatus*
heteros (Gr) different from the usual   *mus* (Gr) a mouse   *niger* (L) black   *caudatus* (L) having a tail. Inhabiting southern Mexico.

Family   CASTORIDAE   2 species
kastōr (Gr) the beaver.

European Beaver   *Castor fiber*
fiber (L) the beaver. Some authorities now consider there is only one species, but there are some small differences between the North American and the Eurasian type. This beaver inhabits Europe and Asia.

Canadian Beaver   *C. canadensis*
-ensis (L) suffix meaning belonging to. In addition to Canada, it is found in the northern part of the USA.

Family   ANOMALURIDAE   9 species, possibly more
anōmalos (Gr) uneven, irregular; here it is taken to mean 'strange' *oura* (Gr) the tail; an allusion to the scales, arranged in two longitudinal rows, on the under part of the basal part of the tail.

Pel's Scaly-tailed Squirrel   *Anomalurus peli*
H. S. Pel was Governor of Dutch Gold Coast, West Africa (now Ghana) from 1840 to 1850. The scaly-tails are squirrel-like animals resembling the flying squirrels; with the exception of *Zenkerella insignis*

(below), they too have a web of skin extending from the forelegs to the hind legs and out to the tail; this enables them to glide from tree to tree for distances of about 20 m (60 to 70 ft). Inhabiting western and central parts of Africa.

Beecroft's Scaly-tailed Squirrel   *Anomalurops beecrofti*
ops from *opsis* (Gr) aspect, appearance; it is very much like *Anomalurus* (above). John Beecroft was an Englishman who was made Governor by the Spaniards of their island Fernando Po, in 1844. Inhabiting western and central parts of Africa.

Pygmy Scaly-tailed Squirrel   *Idiurus zenkeri*
idios (Gr) one's own, private; can mean peculiar, distinct, hence strange   *oura* (Gr) the tail; this is a reference to the thinly haired tail, which has a number of rows of small scales on the under side near the base. G. Zenker (1855–1922) was a botanist and ornithologist who spent several years in West Africa from 1900 onwards. Inhabiting central Africa and Cameroun.

Non-gliding Scaly-tail   *Zenkerella insignis*
-ellus (L) the diminutive suffix which is sometimes used with a personal name to mean 'of Zenker' (see above)   *insignis* (L) remarkable, notable; of the nine species this is the only one that does not have the gliding membrane. Inhabiting Cameroun.

Family   PEDETIDAE   1 species
pēdētēs (Gr) a dancer, a leaper.

Springhaas or Springhare   *Pedetes capensis*
-ensis (L) suffix meaning belonging to; taking its name from the Cape of Good Hope, South Africa. It also inhabits parts of eastern Africa; it has big powerful hind legs and moves by a series of huge leaps.

Suborder   MYOMORPHA

mus (Gr) a mouse   *morphē* (Gr) form, resemblance; 'mouse-like' many are much bigger than a mouse.

Family   CRICETIDAE   about 570 species
cricetus (New L) derived from *criceto* (It) the hamster.

Coues's or Texan Rice Rat   *Oryzomys couesi*
oruza (Gr) rice   *mus* (Gr) a mouse. Dr E. B. Coues (1842–1899) wa

an American surgeon and naturalist. This rice rat inhabits Belize in Central America; it may range northwards to Texas.

Black-eared Rice Rat *O. melanotis*
melas (Gr) black *ous* (Gr), genitive *ōtos*, the ear. Inhabiting Central and South America.

Alfaro's Rice Rat *O. alfaroi*
Dr A. Alfaro (1865–1951) was a zoologist, and at one time was Director of the Natural History Museum in Costa Rica. Inhabiting Central America.

Common Rice Rat *O. palustris*
palustris (L) marshy, boggy; can mean living in marshy places. Inhabiting southern USA, and south-eastern and northern parts of South America.

Jamaican Rice Rat *O. antillarum*
antillarum (New L) of the Antilles, a name applied to the islands of the West Indies, and which includes Jamaica. This rat is now very rare and may be extinct.

American Harvest Mouse *Reithrodontomys humulis*
rheithron (Gr) a stream, also a channel *odous* (Gr), genitive *odontos*, a tooth *mus* (Gr) a mouse; 'groove-toothed mouse'; there are grooves on the upper incisor teeth *humus* (L) the ground, hence *humilis*, on the ground. Inhabiting North America.

White-footed Mouse *Peromyscus maniculatus*
ēra (Gr) a pouch *mus* (Gr) a mouse *-iscus* (L) diminutive suffix; little pouched mouse'; it has cheek pouches like *Perognathus penicillatus*; *manus* (L) a hand, hence *manicula*, a little hand *-atus* (L) suffix meaning provided with. Inhabiting North and Central America.

White-footed Mouse *P. leucopus*
ukos (Gr) white *pous* (Gr) the foot. Range as above.

Cotton Rat *Sigmodon hispidus*
igma (Gr) the letter Σ; *oudous (=odōn)* (Gr) a tooth; an allusion to the sigmoid pattern of the enamel of the molars when their crowns are worn down *hispidus* (L) hairy, shaggy; a reference to the harsh hair. Inhabiting southern USA and the northern part of South America.

Dusky-footed Woodrat   *Neotoma fuscipes*
neos (Gr) new   *tomos* (Gr) sharp, cutting; an allusion to the teeth
indicating a new genus of rodent   *fuscus* (L) dark-coloured   *pes* (L)
the foot. Inhabiting Canada and the coastal strip of western USA.

Bushy-tailed Woodrat   *N. cinerea*
cinis (L) ashes, hence *cinereus*, ash-coloured. Inhabiting British
Columbia and most of the western part of the USA.

Eastern Woodrat or Florida Packrat   *N. floridana*
-anus (L) suffix meaning belonging to. Named from Florida, but also
inhabiting other parts of south-eastern USA.

Fish-eating Rat   *Ichthyomys stolzmanni*
ikhthus (Gr) a fish   *mus* (Gr) a mouse; an allusion to its habit of eating
fish; it is notably modified for a semi-aquatic life;   named after Dr
Jean Stanislas Stolzmann (1854–1928), 'One of the best known and
most successful of Peruvian collectors; the discoverer of many new
mammals' (O. Thomas, 1893). He was Director of the Branicki
Zoological Museum in Warsaw from 1887 until his death in 1928
This rat inhabits Peru.

Dwarf Hamster   *Phodopus sungorus*
phōs (Gr), genitive *phōdos*, a burn, a blister   *pous* (Gr) the foot; the
tubercles on the soles of the feet form a blister-like mass   *sungorus* i
from Dzungaria, a vast valley between the Altai Mountains and the
Tienshan Mountains in Sinkiang, China; the name was given by
Pallas in 1777. It also inhabits Siberia and Manchuria.

Common Hamster   *Cricetus cricetus*
cricetus (New L) derived from *criceto* (It) the hamster. It is found in
Russia, western Asia, Africa and parts of Europe.

Grey Hamster   *Cricetulus triton*
-ulus (L) diminutive suffix; a small hamster   Triton was a Greek sea
god; the word is used here to indicate 'large'; 'A triton among the
minnows'. Although this genus consists of small short-tailed hamster
this species is the largest in the genus. Inhabiting eastern Europe and
northern Asia.

Golden Hamster   *Mesocricetus auratus*
mesos (Gr) middle   *cricetus* (New L) the hamster; indicating its inter
mediate position between *Cricetus* and *Cricetulus* (above)   *auratus* (L
golden. This hamster, well known as a domestic pet, is descende

from a family of hamsters found in Syria in 1930; apart from this occasion it has never been seen in the wild, or at least any sightings have not been recorded.

Maned Rat   *Lophiomys imhausii*
lophos (Gr) a crest   *mus* (Gr) a mouse; sometimes known as the Crested Hamster, it has a crest of erectile hairs along the back which are raised when the animal is alarmed. According to Dr Wilhelm C. H. Peters, 'a skull of the singular rodent lately described by M. Alphonse Milne-Edwards under the name *Lophiomys imhausii*, in the zootomical collection at Berlin, had been obtained by Dr Schweinfurth from the tombs of Maman, northward of Kassalá in Upper Nubia' *imhausii* apparently refers to a Monsieur Imhaus, of Aden, who purchased a specimen that had been collected in north-east Africa; it is reported that this unusual rodent has also been seen in Uganda. Dr Wilhelm C. H. Peters (1815-1884) was a Professor of Zoology in Berlin. He was in East Africa from 1842 to 1848. Professor Alphonse Milne-Edwards (1835-1900) was a French zoologist. Dr G. A. Schweinfurth (1836-1925) was an author and naturalist who explored Africa during the nineteenth century.

Collared Lemming   *Dicrostonyx hudsonius*
dikroos (Gr) forked, cloven   *stonux* (Gr) a sharp point   *onux* (Gr) a claw; in winter two claws on the forefeet become enlarged and prominent; the purpose is not established, but may have something to do with digging in the snow   *hudsonius*, of Hudson; named from Hudson Bay. It is widespread in the Arctic regions.

Norwegian Lemming   *Lemmus lemmus*
lemmus (New L) the lemming, derived from the Norwegian *lemming*. This is the lemming that is occasionally involved in mass migration, which may take them into the sea where many are drowned. Other species live in northern North America and northern Asia.

Bank Vole   *Clethrionomys glareolus glareolus*
(*Clethrionomys* formerly *Evotomys*)
klethron (Gr) a bolt or bar for closing a door   *mus* (Gr) a mouse; it is said this means 'bolt-toothed mouse', possibly because the enamel ridges on the teeth are rounded and without the angular ridges found in the Common Vole. Another reason could be that the molar teeth are rooted, i.e. 'bolted', and so differ from true voles which have constantly growing molars. It has also been suggested that the name

derives from *klethra* (Gr) an alder   *-ion* (Gr) diminutive suffix, and meaning 'alder-grove mouse', from its habitat   *glarea* (L) gravel *-olus* (L) diminutive suffix; actually it usually frequents earth banks rather than gravelly areas. It is widespread in Europe.

Jersey Bank Vole   *C. g. caesarius*
Caesarea is the Roman name for Jersey, one of the Channel Islands. The subspecific name indicates a subspecies of *C. glareolus*. The Bank Voles are sometimes known as Red-backed Voles; the fur on the back is a chestnut-red colour.

Skomer Bank Vole   *C. g. skomerensis*
Skomer is an island off the coast of Pembrokeshire, Wales. A subspecies; see above.

Raasay Bank Vole   *C. g. erica*
erice (L) heath, ling, from *ereikē* (Gr) heath, heather; 'a heath-vole'. Raasay is an island near the Isle of Skye, Inverness-shire, Scotland. A subspecies, see above.

Martino's Snow Vole   *Dolomys bogdanovi*
dolos (Gr) deceit   *mus* (Gr) a mouse; 'unter Anspielung auf die Bedeutung des Namens *Phenacomys*'—evidently on account of the puzzling affinities of the type species (see below). Named after Professor M. N. Bogdanov (1841–1888) of St Petersburg (Leningrad). A fossil of this vole was found in Hungary in 1898, and it was not until 1925 that a living specimen was found. First described by V. and E. Martino, and possibly given the name Snow Vole because it lives in the mountains, as high as 600 m (2,000 ft) in Yugoslavia.

Water Vole   *Arvicola terristris*
arvum (L) ploughed land, a field   *colo* (L) I till, I cultivate; can mean dwell in, inhabit   *terristris* (L) of the ground; it lives on river banks and is a good swimmer. Sometimes known as the Water Rat, it is a vole and not a rat. Widespread in Europe including the British Isles but not Ireland. Also found in the Arctic, in North America, Russia and Siberia, and ranging southwards to parts of Asia Minor and Israel.

Muskrat   *Ondatra zibethica*
Ondatra is a North American Indian name for the muskrat   *zibett* (It) the civet-cat; *zibethicus* (New L) civet-odoured. Muskrats were brought to Great Britain in 1929 and kept in fur-farms for the valuable fur, but many escaped and became established over large areas. The

were then registered as pests and exterminated by about 1940. Inhabiting North America.

Newfoundland Muskrat   *O. obscura*

obscura (L) dark, dusky. The muskrat is an aquatic animal and adapted for swimming; this species is found only on the island of Newfoundland.

Tree Mouse or Heather Vole   *Phenacomys intermedius*

phenax (Gr), genitive *phenakos*, a cheat   *mus* (Gr) a mouse; an allusion to the fact that 'the external appearance of the animal gives no clue to its real affinities'   *intermedius* (L) between, intermediate; probably a reference to the tail which is of medium length; other species have long tails. Inhabiting North America.

Common Field Vole   *Microtus agrestis*

mikros (Gr) small   *ous* (Gr), genitive *ōtos*, the ear   *agrestis* (L) belonging to the fields. A characteristic of the voles is that the ears are small and inconspicuous. Inhabiting Europe and northern Asia.

Steppe Lemming   *Lagurus lagurus*

lagos (Gr) a hare   *oura* (Gr) the tail; it has a short tail, similar to a hare. Inhabiting southern Russia and western Siberia.

Chinese or Yellow Steppe Lemming   *L. luteus*

luteus (L) the colour of the plant lutum, saffron-yellow. It lives in Mongolia and northern China.

American Steppe Lemming or Sagebrush Vole   *L. curtatus*

curtus (L) shortened   *-atus* (L) suffix meaning provided with; probably referring to the short legs and tail. Inhabiting the western part of the USA and Canada.

Pygmy Gerbil or Sand Rat   *Gerbillus gerbillus*

gerbil, and *jerboa*, derived from *jarbū* (Ar) a rodent   *-illus* (L) diminutive suffix; 'a little gerbil'. Inhabiting Palestine and ranging south to northern Africa.

Indian Gerbil   *Tatera indica*

tatera, obscure; Professor Lataste says it is a euphonic name of unknown origin   *Indica* (L) Indian. Inhabiting sandy and semi-desert areas of India, Arabia and Africa. Professor Fernand Lataste was a French zoologist. His *Étude de la Faune des Vertébrés de Barbarie (Algérie, Tunisie et Maroc)*, published in 1885, was for many years a standard work on the animals of North Africa.

Mid-day Jird   *Meriones meridianus*
Mērionēs and Idomeneus were companions in arms in the Trojan War of Greek mythology. Since the early days of classification, Mērionēs has been used as the name of a genus of gerbils, and until fairly recently Idomeneus was the name of a subgenus   *meridianus* (L) at mid-day; there seems no obvious reason for the name. Jird is another name for the gerbil from the Berber name *gherda*. Living in desert areas of northern Africa.

Fat Sand Rat   *Psammomys obesus*
psammos (Gr) sand   *mus* (Gr) a mouse   *obesus* (L) fat; this is a kind of gerbil living in desert areas of eastern Europe, Africa and south-west Asia.

Great Gerbil   *Rhombomys opimus*
rhombos (Gr) a rhombus, a parallelogram having its sides equal and two angles oblique; i.e. lozenge-shaped; a reference to the enamel on the upper molars which shows a lozenge-shaped pattern   *mus* (Gr) a mouse   *opimus* (L) rich, fat. Inhabiting the deserts of Mongolia Turkestan and Iran.

Family   SPALACIDAE   3 species
spalax (Gr), genitive *spalakos*, a mole.

Palestine Mole Rat   *Spalax ehrenbergi*
Dr C. G. Ehrenberg (1795–1876) worked in North Africa in 1820 where these mole rats live; they range north to Palestine and east to Russia.

Family   RHIZOMYIDAE   18 species
rhizōma (Gr) a root   *mus* (Gr) a mouse; they dig in the ground for roots, their main food.

African Mole Rat   *Tachyoryctes splendens*
takhus (Gr) fast, swift   *orussō* (Gr) I dig   *oruktēs* (Gr) one who digs 'a fast digger'   *splendeo* (L) I shine; some have a shining black coat. They are very fast diggers, and eat roots and possibly cane. Inhabiting East Africa.

Family   MURIDAE   about 460 species
mus (L), genitive *muris*, a mouse or rat.

Marmoset Mouse *Hapalomys longicaudatus*
hapalos (Gr) soft *mus* (Gr) a mouse; this refers to the soft fur *longus* (L) long *cauda* (L) the tail of an animal *-atus* (L) suffix meaning provided with. Inhabiting the Malay Peninsula, Thailand and Indo-China.

Harvest Mouse *Micromys minutus*
mikros (Gr) small *mus* (Gr) a mouse *minutus* (L) small, minute; it only weighs about 7 g ($\frac{1}{4}$ oz). Inhabiting Europe, including the British Isles, eastern Europe and China.

Long-tailed Field or Wood Mouse *Apodemus sylvaticus*
apodemus (Gr) away from home; in the fields; the name is given to distinguish it from the House Mouse *silva* (L) a wood, hence *silvaticus*, of woods or trees; applied to plants and animals it usually means wild. Inhabiting Europe, North Africa and parts of Asia.

Yellow-necked Mouse *A. flavicollis*
flavus (L) yellow *collum* (L) the neck. The range is similar to the Long-tailed Mouse, above.

New Guinea Giant Rat *Hyomys goliath*
us (Gr), genitive *huos*, a pig *mus* (Gr) a mouse Goliath was the Philistine giant in the Bible. This large rat, with a massive skull, can be up to 76 cm (30 in) long, including the tail.

Striped Grass Mouse *Lemniscomys striatus*
lemniscus (L) a ribbon; in this case the 'ribbons' are stripes *mus* (L) a mouse *stria* (L) a groove, a furrow, hence *striatus* (New L) striped; it has buff stripes on a dark brown coat. Living in Africa.

Zebra Mouse *Rhabdomys pumilio*
rhabdos (Gr) a rod, a stick, so *rhabdōtos* (Gr) streaked, striped *mus* (Gr) mouse; it has several light and dark stripes along the centre of the back *pumilio* (L) a dwarf; sometimes known as the Four-striped Rat, is only about the size of a house mouse. It lives in Uganda and Kenya and ranges southwards into South Africa.

Black Rat *Rattus rattus* (*Rattus* formerly *Epimys*)
rattus (New L) a rat; sometimes known as the Ship Rat. This rat, and the Brown Rat (below), probably originated in Asia, but can now be found almost anywhere in the world, except the Arctic and Antarctic regions.

Brown or Norwegian Rat   *R. norvegicus*
-icus (L) suffix meaning belonging to; of Norway; it does not originate
in Norway and is not confined to that country. Like the Black Rat it
probably originated in Asia. Now found throughout the world, and
probably even more common in the British Isles than the Black Rat.

Multimammate Rat   *Mastomys natalensis*
mastos (Gr) the breast   *mus* (Gr) a mouse   *-ensis* (L) suffix meaning
belonging to; found in Natal, South Africa, in 1843, it inhabits most
of Africa except desert areas. Multimammate refers to the fact that
it may have as many as sixteen pairs of teats.

Stick-nest Rat   *Leporillus apicalis*
lepus (L), genitive *leporis*, a hare   *-illus* (L) diminutive suffix; with its
large ears it is rather like a small hare   *apex* (L), genitive *apicis*, the
top, the tip   *-alis* (L) suffix meaning relating to; it has a white tip to
the tail. It builds large nests made of sticks. Inhabiting southern
Australia.

Giant Naked-tailed Rat   *Uromys caudimaculatus*
oura (Gr) the tail   *mus* (Gr) a mouse   *cauda* (L) the tail of an animal
macula (L) a spot   *-atus* (L) provided with; 'spotted tail'; the tail i
of two colours, and is hairless. A very big rat inhabiting New Guinea

House Mouse   *Mus musculus*
mus (L) a mouse or a rat; under this name the Romans also included
animals like the marten and sable   *-culus* (L) diminutive suffix
'little mouse'. It is found throughout the world wherever there ar
people living, even in Arctic regions.

New Guinea Kangaroo Mouse   *Lorentzimys nouhuysii*
Dr H. A. Lorentz (1871–1944) was in the Dutch Consular Service
in New Guinea during the period 1903 to 1910. Captain J. W. \
Nouhuys (born 1869) of the Dutch Navy was also there at that time
This mouse has strong hind legs and hops when moving about.

Australian Kangaroo Mouse   *Notomys mitchelli*
notos (Gr) the south   *mus* (Gr) a mouse   Sir Thomas Livingston
Mitchell (1792–1855) was a Scottish surveyor and explorer, and wa
the Surveyor General for New South Wales in 1827. Sometime
known as Hopping Mice, they hop rather than walk or run; they a
in no way related to Kangaroos.

Fawn-coloured Kangaroo Mouse   *N. cervinus*
cervus (L) a stag, a deer, hence *cervinus*, tawny, like a deer. See notes above; inhabiting Australia.

Cairo Spiny Mouse   *Acomys cahirinus*
akōkē (Gr) a sharp point   *mus* (Gr) a mouse; for self protection the hair on the body of this mouse has evolved into spiny prickles. When Cairo was founded in AD 968 it was named El-Kahira, meaning 'The Victorious'; the name was gradually corrupted and became Cairo. In addition to Africa, it inhabits southern Asia.

Cape Spiny Mouse   *A. subspinosus*
sub- (L) prefix meaning under; can also mean somewhat, slightly *spina* (L) a thorn; spinosus, thorny, full of thorns; it is less prickly than other spiny mice. Inhabiting South Africa.

Bandicoot Rat or Pig Rat   *Bandicota indica*
bandikokku (Telugu) a pig rat   Telegu is a language spoken in the central and eastern parts of southern India   *Indica* (L) Indian. Sometimes known as the Malabar Rat, Malabar being a district in southwest India. It is supposed to smell rather like a pig, and grunts like a pig at times. Inhabiting southern India.

Bengal Bandicoot Rat   *B. bengalensis*
ensis (L) suffix meaning belonging to. Sometimes known as a Mole Rat, it is a good burrower, like a mole. First found in the Bengal area, it is widespread in southern Asia, including the Malay Peninsula, Sumatra and Java.

Giant Rat or Hamster Rat   *Cricetomys gambianus*
cricetus (New L) from *criceto* (It) the hamster   *mus* (Gr) a mouse   also *mus* (L) a mouse or a rat   *-anus* (L) suffix meaning belonging to; 'of Gambia'. It is not a hamster, but a very large rat, sometimes measuring more than 75 cm (30 in) overall. Inhabiting forested tropical areas of West Africa.

African Climbing or Tree Mouse   *Dendromus mesomelas*
dendron (Gr) a tree   *mus* (Gr) a mouse   *mesos* (Gr) middle   *melas* (Gr) black; it has a black stripe down the middle of the back. Widespread in Africa.

South African Swamp Rat   *Otomys irroratus*
ōs (Gr), genitive *ōtos*, the ear   *mus* (Gr) a mouse; T. S. Palmer does

not give any explanation for 'ear-mouse' *irroratus* (L) moistened with dew, wetted; a reference to its damp marshy habitat. Inhabiting South Africa.

New Guinea Rat   *Mallomys rothschildi*
mallos (Gr) wool, a fleece   *mus* (Gr) a mouse; it has thick woolly fur; named after L. W. Rothschild, 2nd Baron, FRS (1868-1937) who founded the Zoological Museum, Tring, England in 1889; he also wrote much on zoology. Another large rat, up to 86 cm (34 in) from nose to tail; living in New Guinea.

Philippine Cloud Rat   *Phloeomys cumingi*
phloios (Gr) bark of trees, peel   *mus* (Gr) a mouse; 'suggested by the habit of the animal, which Mr Cuming states feeds chiefly on the bark of trees'. H. Cuming (1791-1865) was a sailmaker in Valparaiso, Chile, in 1819. He collected the first specimen of this gigantic rat in Luzon, in the Philippines. It is probably named Cloud Rat because of its mountain habitat. It is a very large mountain species inhabiting the Philippines.

Bushy-tailed Rat   *Crateromys schadenbergi*
krateros (Gr) strong, mighty   *mus* (Gr) a mouse; referring to the fact that it is about the largest and heaviest member of the Muridae Named after Dr A. Schadenberg who sent the first specimens to Dresden Museum in 1894. He collected them on Mount Data, Luzon in the Philippines. It is bushy like a Persian cat, the body as well as the tail and lives in the Philippine mountains.

Philippine Shrew Rat   *Rhynchomys soricoides*
rhunkhos (Gr) the snout, beak   *mus* (Gr) a mouse; referring to the very long snout   *sorex* (L) genitive soricis, a shrew-mouse   *-oides* (New L) from *eidos* (Gr) apparent shape, resemblance; the pointed snout is like that of a shrew. Inhabiting the Philippines.

Beaver Rat   *Hydromys chrysogaster*
hudōr (Gr) water; in composition the prefix *hudro-* is used   *mus* (Gr) a mouse   *khrusos* (Gr) gold   *gastēr* (Gr) the stomach; 'golden-bellied water-rat'. A good swimmer, and very common in New Guinea and Australia.

Shaw Mayer's Mouse   *Mayermys ellermani*
F. W. Shaw Mayer (born 1899) an Australian zoologist. He collected for the British Museum (Natural History) and in New Guinea in 19

for the Zoological Museum at Tring. Sir John Ellerman (1909-1973) was the son of Sir John Reeves Ellerman, founder of the Ellerman Steamship Line. He was the author of a standard work on rodents. An interesting rodent, having only one molar tooth on each side of upper and lower jaws; this is unique among rodents. It lives in New Guinea.

Family GLIRIDAE about 28 species

glis (L), genitive *gliris*, a dormouse; the English name probably comes from *dormio* (L) I sleep; they have a very long period of hiberation, and so have earned a reputation for being sleepy.

Edible or Fat Dormouse *Glis glis* (*Glis* formerly *Myoxus*)

It used to be eaten by the Romans; it becomes very fat before hibernating. It inhabits the southern part of Europe, Russia and ranges south to Syria. It has been introduced into the southern Midlands in England.

Hazel or Common Dormouse *Muscardinus avellanarius*

Muscardinus (New L) from *muscardin* (Fr) a doormouse, from *muscadin*, a musk-scented lozenge; a reference to the odour of the animal *avellana* or *abellana* (L) filbert or hazel nut, from Abella, a town in Campania, Italy, abounding in fruit-trees and nuts *-arius* (L) belonging to; these nuts are the favourite food of this dormouse. Inhabiting Europe and Asia.

Garden Dormouse *Eliomys quercinus*

eleios (Gr) a kind of dormouse *mus* (Gr) a mouse *quercus* (L) the oak *-inus* (L) suffix meaning pertaining to; dormice like acorns and so would often be seen in oak trees; they also make nests in trees. Inhabiting Europe, north-west Africa and south-west Asia.

Forest Dormouse *Dryomys nitedula*

drus (Gr), genitive *druos*, the oak (see above) *mus* (Gr) a mouse *nitedula* (L) a small mouse, or a dormouse. From Europe and Asia.

Japanese Dormouse *Glirulus japonicus*

glis (L), genitive *gliris*, a dormouse *-ulus* (L) diminutive suffix *-icus* (L) belonging to; 'little dormouse of Japan'.

Mouse-tailed Dormouse *Myomimus personatus*

mus (Gr), genitive *muos*, a mouse *mimos* (Gr) an actor, a mimic *persona* (L) a mask, as for drama *-atus* (L) suffix meaning provided with; unlike other dormice this one does not have a bushy tail; 'one who imitates an ordinary mouse'. This is a rare species inhabiting the Russo-Iranian border, and surprisingly has recently been found in Bulgaria.

African Dormouse   *Graphiurus murinus*
grapheion (Gr) a pencil   *oura* (Gr) the tail; the tail has a pencil of hairs at the tip   *murinus* (L) mouselike. Inhabiting forested areas throughout the whole of Africa.

Family   PLATACANTHOMYIDAE   2 species
platus (Gr) wide, flat   *akantha* (Gr) a thorn, a prickle   *mus* (Gr) a mouse; they have flattened spines in the hair and are known as 'spiny dormice' because the hairs are sharp and prickly.

Spiny Dormouse   *Platacanthomys lasiurus*
lasios (Gr) hairy   *oura* (Gr) the tail; the tail is scaly at the base and has a bushy tip. Inhabiting southern India.

Family   SELEVINIIDAE   1 species

Desert Dormouse   *Selevinia betpakdalaensis*
This animal was first discovered in Central Kazakhstan, a southern desert region of the USSR, by a zoologist named W. A. Selevin, about the year 1938. It is rare and unique, and has to be placed in a family on its own. The name has also been given as *Selevinia paradoxa*, presumably because classifying it was something of a paradox. By the law of priority the original name is the only legitimate one. Bet-Pak-Dala is the name of a village in Kazakhstan close to where the animal was found.

Family   ZAPODIDAE   11 species
za- (Gr) a prefix with intensive meaning; much or very   *pous* (Gr), genitive *podos*, the foot; taken to mean 'strong feet' or 'big feet', as they have strong well developed back legs for jumping.

Northern Birch Mouse   *Sicista betulina*
Sikistan is a Tartar name meaning 'gregarious mouse'   *betula* (L) the birch   *-inus* (L) suffix meaning belonging to; 'of birch trees'. Inhabiting Scandinavia and Finland.

Southern Birch Mouse   *S. subtilis*
subtilis (L) slender, fine, not thick or coarse; could be interpreted as 'graceful'. Inhabiting southern Russia and Rumania.

Jumping Mouse   *Zapus hudsonius*
Zapus, see above. Named from Hudson Bay, Canada; it inhabits other areas of North America.

Szechuan Jumping Mouse   *Eozapus setchuanus*
ēōs (Gr) the dawn; can also mean the east; meaning living in the east *zapus*, as above. Szechwan is a province in China. Inhabiting eastern Asia.

Family   DIPODIDAE   25 species, possibly more
di- from *dis-* (Gr) two   *pous* (Gr), genitive *podos*, the foot; they appear to have only two feet because the front legs are tiny and not used for moving about; they hop like kangaroos and the hind legs are long and well developed. Jerboa is from *jarbū* (Ar) a rodent.

Feather-footed Jerboa   *Dipus sagitta*
Dipus, see above   *sagitta* (L) an arrow; this suggests 'feather-tailed', referring to the tuft of hair at the tip of the tail, like the feathers on an arrow; it also has hairy feet. Inhabiting desert areas of Africa and Asia.

Egyptian Jerboa   *Jaculus jaculus*
iaculor (= *iaculor*) (L) to throw a javelin, hence *iaculus*, thrown, darting; it is only 12–18 cm (5–6 in) long but can leap a distance of nearly 2 m (6 ft). Inhabiting desert areas of North Africa.

Euphrates Jerboa   *Allactaga euphratica*
alak-dagha is a Mongolian name for the jerboa   *-icus* (L) suffix meaning belonging to; 'of the Euphrates'. It inhabits the desert lands of Arabia.

Siberian Jerboa   *A. sibirica*
Inhabiting Siberia; this usually means the whole territory between the Ural Mountains and the Pacific Ocean.

Three-toed Dwarf Jerboa   *Salpingotus kozlovi*
salpinx (Gr), genitive *salpingos*, a war trumpet   *ous* (Gr), genitive *ōtos*, the ear; it has funnel-shaped or trumpet-shaped ears; named after General P. K. Kozlov (1863–1935) a Russian zoologist who explored central Asia during the years 1899 to 1926. This remarkable little jerboa, measuring only 5 cm (2 in) head and body, has hairy tufts

under the three hind toes. Named only recently, in 1922, it inhabits the Gobi Desert in Mongolia.

Long-eared Jerboa   *Euchoreutes naso*
eu- (Gr) prefix meaning well   *khoreutēs* (Gr) a dancer; 'a good dancer'; a reference to the animal's mode of progression by leaps   *nasus* (L) the nose; it has a pointed nose and very long ears. Inhabiting Sikiang and Inner Mongolia.

Suborder   HYSTRICOMORPHA
hustrix (Gr), genitive *hustrikhos*, a hedgehog, a porcupine   *morphē* (Gr) shape, resemblance.

Family   HYSTRICIDAE   15 species

Great Crested Porcupine   *Hystrix cristata*
cristatus (L) crested. Inhabiting southern Europe, North and West Africa and south-western Asia.

Malayan Porcupine   *H. brachyura*
brakhus (Gr) short   *oura* (Gr) the tail. Inhabiting Malaya, Borneo and Sumatra.

Brush-tailed Porcupine   *Atherurus africanus*
athēr (Gr) the beard of an ear of wheat   *oura* (Gr) the tail; it has a tuft of hair on the tip of the tail. Inhabiting West Africa and part of western central Africa.

Rat Porcupine   *Trichys lipura*
thrix (Gr), genitive *thrikhos*, hair; the body has a lot of stiff hair mixed with spines and bristles   *leipō* (Gr) I am lacking, wanting   *oura* (Gr) the tail; the tail is shorter than in the allied genus *Atherurus*. Inhabiting the southern part of Malaysia, and Borneo and Sumatra.

Family   ERETHIZONTIDAE   11 species, possibly more
erethizō (Gr) I rouse to anger, I irritate.

Canadian Porcupine   *Erethizon dorsatum*
dorsum (L) the back   *-atus* (L) suffix meaning provided with; 'having a back that irritates'. Inhabiting large areas of northern North America, including Canada.

Brazilian Tree Porcupine   *Coendou prehensilis*
Coendou is a Brazilian native name for the porcupine   *prehenso* (L

I lay hold of -*ilis* (L) adjectival suffix denoting capability; 'able to hold'; it has a prehensile tail used in tree climbing. Inhabiting Central and South America.

Family CAVIIDAE 15 species
cavia from *çaviá* (Port) now *savia* from the Tupi word *sawiya*, a rat.

Wild Guinea Pig or Restless Cavy *Cavia porcellus*
porcus (L) a pig, hence *porcellus*, a little pig; it is no relation to the pig family, the Suidae, nor does it come from Guinea; a mistake for Guyana. Inhabiting South America.

Brazilian Cavy *C. aperea*
aper (L) a wild boar, or pig.

Rock Cavy or Moco *Kerodon rupestris*
keras (Gr) the horn of an animal *odōn* (Gr Ionic dialect) a tooth; the reason is obscure; T. S. Palmer offers no explanation in his standard work *Index Generum Mammalium* (1904) *rupes* (L) a rock, hence *rupestris*, living among rocks. Moco is a Tupi name for this cavy; it lives in Brazil.

Patagonian Hare or Mara *Dolichotis patagona*
dolikhos (Gr) long *ous* (Gr), genitive *ōtos*, the ear; it is not a hare, but has long legs and rather bigger ears than other members of the family *mara* is American Spanish. Patagonia is a region in the southern part of Argentina.

Salt Desert Cavy *Pediolagus salinicola*
pedion (Gr) a plain, level country *lagōs* (Gr) a hare *sal* (L), genitive *salis*, salt, hence *salinus* (New L) salty *colo* (L) I inhabit; it is not known except in the salt desert areas of the southern part of South America. Some now give *Pediolagus* as a subgenus of *Dolichotis*.

Family HYDROCHOERIDAE 2 species
udōr (Gr) water; in composition the prefix *hudro-* is used *khoiros* (Gr) a young pig.

Capybara or Carpincho *Hydrochoerus hydrochaeris*
Sometimes known as the Water Hog, it is not a hog, but always lives near water and is a good swimmer and diver. It is the largest rodent and can weigh up to 45 kg (100 lb) and measure over 1 m (4 ft) in

length. Capybara and Carpincho are South American names derived from Tupi. Inhabiting a large area of South America except the very southern parts.

Panama Capybara   *H. isthmius*
isthmos (Gr) a narrow passage, also a neck of land between two seas; this capybara lives in the Panama area, so the name refers to the Isthmus of Panama.

Family   DINOMYIDAE   1 species
deinos (Gr) terrible, formidable   *mus* (Gr) a mouse; it has the appearance of an enormous guinea pig, with a head and body length of up to 76 cm (30 in) and a tail of 20 cm (8 in).

Pacarana or False Paca   *Dinomys branickii*
Named after Grafen Constantin Branicki. Jelski made zoological journeys of discovery in Surinam and Peru which were rich in results; he was able to do this owing to the splendid liberal support of Branicki. Professor K. Jelski (1838–1896) was a Polish zoologist and Curator of the Museum in Cracow from 1878 until his death in 1896. Grafen Constantin Branicki (1823–1884), after whom the animal was named, was a wealthy Polish nobleman whose son and nephew established the Branicki Zoological Museum in Warsaw in 1887, Dr J. S. Stolzmann being the Director (see page 124). *Paca* (Sp from Tupi) a South American rodent   *rana* (Tupi) false; so named because it is not a true paca of the family Dasyproctidae (below). It inhabits mountainous areas in Peru.

Family   DASYPROCTIDAE   17 species
dasus (Gr) hairy   *proktos* (Gr) the hindpart, the rump; the hair is not confined to the rump though it is longer there and usually of a different colour which makes it conspicuous.

Paca   *Cuniculus paca*   (*Cuniculus* formerly *Coelogenys*)
cuniculus (L) a rabbit; it is not a rabbit but more like a very large rat with a short tail and a spotted coat   *paca* is a South American Spanish name derived from Tupi. Inhabiting a large area from Mexico south to the Argentine.

Mountain Paca   *C. taczanowskii*
Dr W. Taczanowski (1819–1890) was a Polish zoologist who made a

expedition to Peru in 1884. This paca lives high in the Andes Mountains, possibly up to about 3,000 m (10,000 ft).

Orange-rumped Agouti *Dasyprocta aguti*
Aguti, or *acuti*, is a South American Spanish name; this Agouti lives in Brazil.

Acouchi *Myoprocta acouchy*
mus (Gr), genitive *muos*, a mouse *proktos* (Gr) hinder parts, the rump (from *Dasyprocta*, 'hairy buttocks', see above); the acouchi does not have the coloured rump hair of the agouti, but the hair is thick, and erected when fighting: acouchy is a native name. Inhabiting the northwestern part of South America.

Family CHINCHILLIDAE 6 species
Chinchilla is really a Spanish name for this animal, probably derived from the Quechua Indian.

Chincilla *Chinchilla laniger*
lana (L) wool *gero* (L) I carry; the coat is possibly the best and most highly valued of any animal; it is very soft and silky. Inhabiting the mountains in northern Chile.

Mountain Viscacha *Lagidium peruanum*
lagos (Gr) a hare *-idium* (New L) derived from *-idion* (Gr) a diminutive suffix *peruanum*, of Peru; it lives in the mountains of Peru. *viscacha* (New L) from American-Spanish *vizcacha*.

Plains Viscacha *Lagostomus maximus*
lagos (Gr) a hare *stoma* (Gr) the mouth; 'the hare-mouthed one' *maximus* (L) the largest; a much larger animal than the chincilla. Inhabiting the southern part of South America.

Family CAPROMYIDAE 11 species
capros (Gr) a wild boar *mus* (Gr) a mouse; Professor A. G. Desmarest (1784-1838) a French Professor of Zoology said the name was a result of the animal's resemblance to a wild boar in general appearance, character of hair, colour and manner of running.

Hutiacouga *Capromys pilorides*
Known locally as the Pilori Rat, the name is probably of Arawak origin, an Indian people of South America formerly living in Cuba,

where this hutia lives   -*ides* (L) suffix indicating a relationship   *hutia* is a Spanish name for a rodent   *couga*, probably from *cougar*, a Guarani name for a type of cat. Inhabiting Cuba.

Hutiacarabali   *C. prehensilis*
prehenso (L) I take hold of   -*ilis* (L) suffix denoting capability; 'able to hold'; it has a prehensile tail used in tree climbing   *carabali*, obscure. Inhabiting Cuba.

Jamaican or Short-tailed Hutia   *Geocapromys brownii*
gē (Gr) the earth, ground; a reference to its ground-dwelling habits, unlike other species of *Capromys* which live in trees; named after Patrick Browne (1720-1790) from whose *Civil and Natural History of Jamaica* J. B. Fischer (1730-1793) a German zoologist, took the account of this Hutia or Indian Coney. This animal is rather like a very large rat with no tail, for this is only a stump and hidden under the fur. It lives in the Blue Mountains area of Jamaica.

Dominican House Rat   *Plagiodontia aedium*
plagios (Gr) slanting, oblique   *odous* (Gr), genitive *odontos*, a tooth a reference to the diagonal grooves in the upper molars   *aedes* (L) a room, a house   -*ium* (L) a suffix sometimes used to denote a place as in 'aquarium'. Inhabiting Haiti, West Indies.

Coypu   *Myocastor coypus*
mus (Gr) a mouse   *kastōr* (Gr) a beaver; it is largely aquatic and lives on river banks, but it is not a beaver   *coypu* was originally a South American native name. Valued for its fur, known as nutria, it was brought to England and reared in captivity. Many animals escaped and became a threat to agriculture in East Anglia. However it is gradually being exterminated. Originally inhabiting southern South America.

Family   OCTODONTIDAE   8 species
octo (L) eight   *odous* (Gr), genitive *odontos*, a tooth; the grinding surfaces of the lower molars are shaped like a figure eight.

South American Bush Rat or Degu   *Octodon degus*
Degu is a South American native name for this rodent which looks like a large rat; it lives in the mountains of Peru and Chile.

Cururo   *Spalacopus cyanus*
spalax (Gr), genitive *spalakos*, a mole   *pous* (Gr) a foot; it is a burrow

ing animal with feet adapted for this purpose *kuanos* (Gr) dark-coloured, bluish *cururo* is Spanish derived from the Araucan name *curi*; the Araucanians are an Indian people living in southern Chile and adjacent regions of Argentina. This animal looks rather like the Tucotuco (below); it inhabits South America.

Family CTENOMYIDAE 27 species

kteis (Gr), genitive *ktenos*, a rake or comb *mus* (Gr) a mouse; referring to the large fringes on the long claws of the hind feet, probably used to remove dirt from the fur and general grooming.

Tucotuco *Ctenomys brasiliensis*

Tuco-tuco (Sp) an imitation of its cry *-ensis* (L) suffix meaning belonging to; in Spanish Brazil is spelt 'Brasil'. A small greyish-brown burrowing animal up to 30 cm (12 in) long and inhabiting the southern part of South America.

Peruvian Tucotuco *C. peruanus*

-anus (L) suffix meaning belonging to.

Banded Tucotuco *C. torquatus*

torquatus (L) wearing a collar. Inhabiting southern Peru and ranging south to Tierra del Fuego.

Family ABROCOMIDAE 2 species

abros (Gr) soft, luxurious *komē* (Gr) hair; they have dense underfur which is extremely soft and silky.

Rat Chinchilla *Abrocoma bennetti*

Named after E. T. Bennett (1797–1836) who was Secretary of the Zoological Society of London in 1831. It is not a rat, but something between the degu (page 140) and the chinchilla; it lives in Bolivia.

Family ECHIMYIDAE 43 species, possibly more

chi (New L) derived from *ekhinos* (Gr) a hedgehog *mus* (Gr) a mouse; they have sharp bristly fur.

Trinidad Spiny Rat *Proechimys trinitatis*

pro (Gr) before, and *echimys* (see below) *pro-* is used here to denote an allied form *trinitatis*, 'of Trinidad'. Inhabiting South America, but now very rare.

Porcupine Rat   *Euryzygomatomys spinosus*
eurus (Gr) wide   *zugon* (Gr) a yoke, a joining together; the zygomatic
bone is situated in the upper part of the face, and forms the prominence
of the cheek   *mus* (Gr) a mouse; this could be interpreted as 'wide-
cheeked mouse'; it has a broad zygoma   *spina* (L) a thorn, so *spinosus*,
full of thorns. Inhabiting Brazil and Paraguay.

Spiny Rat   *Cercomys cunicularis*
kerkos (Gr) the tail   *mus* (Gr) a mouse; it has a rat-like tail, though
some species have no tail   *cuniculus* (L) a rabbit   *-aris* (L) suffix
meaning pertaining to. Widespread in South America.

Arboreal White-faced Spiny Rat   *Echimys chrysurus*
echi (New L) derived from *ekhinos* (Gr) a hedgehog   *mus* (Gr) a mouse;
it has sharp bristly fur   *khrusos* (Gr) gold   *oura* (Gr) the tail. A tree-
climbing rat with a white face and a pale yellow end to the tail.
Inhabiting north-eastern South America.

Family   THRYONOMYIDAE   2 species
thruon (Gr) a reed   *mus* (Gr) a mouse; they often inhabit reed beds.

Cane Rat or Cutting Grass   *Thryonomys swinderianus*
This cane rat was obtained by the original describer Temminck from
Professor Van Swinderen of Groningen, Netherlands. It is not a rat
or a mouse, but more like a coypu. It is not noticeably fond of cane,
but generally a vegetable grazer, hence 'cutting grass'. Inhabiting
Africa south of the Sahara.

Family   PETROMYIDAE   1 species
petra (Gr) a rock   *mus* (Gr) a mouse; it inhabits rocky areas and sleep
in holes in rocks.

Rock Rat   *Petromus typicus*
tupos (Gr) a blow, the mark of a blow, an impression, like a wax seal
hence *tupikos* (Gr) a type. A rare rat-like animal found only in the
southern part of Africa.

Family   BATHYERGIDAE   16 species, possibly more
bathus (Gr) deep   *ergō* (Gr) I work; they burrow and live almos
entirely underground.

Common Mole Rat *Cryptomys hottentotus hottentotus*

kruptos (Gr) secret, hidden *mus* (Gr) a mouse; seldom seen as they spend most of their life underground; the Hottentots are a native people of Namibia (South West Africa). The name is derived from the Dutch for 'stutterer', on account of the peculiar Hottentot language. This mole-like animal inhabits the western area of Cape Province.

Damaraland Mole Rat or Blesmol *C. h. damarensis*

-ensis (L) suffix meaning belonging to; it inhabits Damaraland, Namibia (South West Africa) and ranges eastwards to Rhodesia. The name 'blesmol' comes from the Dutch *bles*, a blaze, and mole; there is often a white patch on the head.

Lugard's Mole Rat *C. h. lugardi*

Named after F. J. D. Lugard, 1st Baron (1858–1945), a British soldier and colonial administrator in Africa. Lugard and Speke (below) have also had Nile steamers named after them, *SS Lugard* and *SS Speke*. This mole rat inhabits eastern areas of Africa south of the Sahara.

Cape Mole Rat *Bathyergus suillus*

Bathyergus (see above under Family) *sus* (L), genitive *suis*, a pig *illus* (L) diminutive suffix, 'a small pig'. Inhabiting South Africa.

Naked Mole Rat or Sand Puppy *Heterocephalus glaber*

heteros (Gr) different *kephalē* (Gr) the head; probably a reference to the bald head *glaber* (L) bald; the head and body are almost completely hairless. It lives in the hot Somaliland deserts in eastern Africa.

Family CTENODACTYLIDAE 4 species

kteis (Gr), genitive *ktenos*, a rake, a comb *daktulos* (Gr) a finger, a toe; the inner two toes have horny combs on them used for grooming the fur.

Gundi *Ctenodactylus gundi*

gundi is Arabic and probably derived from Berber. Inhabiting the northern part of Africa on the northern borders of the Sahara.

Speke's Pectinator *Pectinator spekei*

pecto (L) I comb, hence *pectinator*, one who combs (see above) Captain J. H. Speke (1827–1864) was the well-known explorer who discovered Lake Victoria and the source of the Nile in 1857. This gundi lives on the borders of the Somali Desert in eastern Africa.

16   Whales, Dolphins and Porpoises
CETACEA

This group consists of the whales, the dolphins and the porpoises. Strictly speaking, they are all whales, the dolphins and the porpoises being small whales. They are mammals that have become adapted to a life in the sea or rivers and have some remarkable features. Like all mammals they are warm-blooded, breath air, and suckle their young with the mother's milk.

The big examples, like the Blue Whale, can weigh well over 100 tonnes; the mouth is enormous and the tongue alone may weigh as much as 2 tonnes! Certainly whales are the largest living creatures, while at the other end of the scale the dolphins and porpoises may be only about 2 m (6 ft) long and weigh less than 45 kg (100 lb).

For purposes of classification the Order Cetacea is divided into two suborders: Mysticeti, the 'moustached whales' and Odontoceti, the 'toothed whales'.

Subclass   EUTHERIA
(see pages 32 and 51)

Order   CETACEA
cetus (L) a large sea creature, the whale or dolphin.

Suborder   MYSTICETI
mustax (Gr), genitive *mustakos*, a moustache   *kētos* (Gr) (= *cetus* (L)) a sea monster, a whale; this refers to the sheets of whalebone, or baleen plates, that hang down from the upper jaw, used for 'netting' the small marine creatures which are their food.

Family   BALAENIDAE   3 species
balaena (L) a whale.

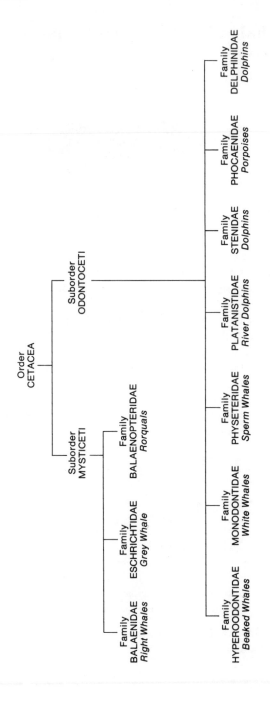

Greenland Right Whale or Bowhead *Balaena mysticetus*
Supposedly called 'right whales' because they were originally the right whales to hunt, as they did not sink when killed. Modern techniques enable other kinds of whales to be caught, though whaling is now controlled. Also known as the Arctic Right Whale; 'bowhead' is suggested by the shape of the enormous head.

Black Right Whale *Eubalaena glacialis*
eu- (Gr) prefix meaning well, nicely; in this case used to indicate 'typical' *balaena* (L) a whale *glacialis* (L) icy. The Right Whales are not confined to Arctic seas and are found in other areas including southern oceans.

Pygmy Right Whales *Caperea marginata*
(*Caperea* formerly *Neobalaena*)
capero (L) I am wrinkled; referring to the wrinkled appearance of the tympanic bone in the skull *margo* (L), genitive *marginis*, the edge; it has dark margins on the baleen plates. Inhabiting southern seas around New Zealand and Australia.

Family ESCHRICHTIDAE 1 species
(formerly RHACHIANECTIDAE)
Named in honour of Daniel Fredrik Eschricht (1798–1863), a Dutch zoologist, and author of several important papers on cetaceans.

Grey Whale or Devilfish *Eschrichtius gibbosus*
(formerly *Rhachianectes*)
gibbus (L) a hump, hence *gibbosus*, humped; in place of the dorsal fin there is a row of nine or ten small humps. Sometimes known as the Californian Whale, it lives in the northern Pacific and migrates to southern seas in the winter. It is the only known species.

Family BALAENOPTERIDAE 6 species
balaena (L) a whale *pteron* (Gr) feathers or wings, or can mean a fin; in allusion to the strong dorsal fin.

Lesser Rorqual *Balaenoptera acutorostrata*
acutus (L) sharp, pointed *rostrum* (L) the beak, snout; it has a more pointed snout than other whales. The name rorqual is derived from the Norwegian *royrkval*, meaning 'red whale'. However, the rorquals are usually bluish-black or grey. Widespread throughout the world, but mostly in Arctic or Antarctic seas.

Rudolphi's or Sei Whale *B. borealis*
boreus (L) northern *-alis* (L) suffix meaning pertaining to. Professor
K. A. Rudolphi (1771–1832) was a professor of anatomy in Berlin
sei is derived from *seje*, the name of a fish living off the coast of Norway
which forms part of this whale's diet. World-wide, moving to warmer
waters in winter for breeding.

Common Rorqual or Fin Whale *B. physalus*
phusalis (Gr) a pipe, a wind instrument; this refers to the blow-hole
through which the whale breathes, in some cases it can make whistling
sounds *rorqual*, see above: it has a high dorsal fin. Widespread
throughout the world.

Blue Whale or Sibbald's Whale *B. musculus*
musculus (L) a little mouse; a word used by Pliny for a sea animal and
Linnaeus thought it was the blue whale, but more probably Pliny
meant it for a pilot-fish that was supposed to guide whales. Although
known as the Blue Whale, it is only a bluish grey, and can even be
grey to light grey and white underneath. It is the largest living creature
and can weigh over 100 tonnes. Sir Robert Sibbald (1641–1722) was
a Professor of Medicine in Edinburgh, and a plant genus *Sibbaldia* is
named after him. This whale is found throughout the world but tends
to inhabit the Arctic and Antarctic seas, migrating to warmer seas in
the winter.

Humpback Whale *Megaptera novaeangliae*
megas (Gr) big *pteron* (Gr) feathers, wings or a fin; it has unusually
long flippers which can be as much as one-third of the body length
novaeangliae of New England; the whale from which this specie
received its name was found on the coast of New England. This is tha
part of the USA comprising Maine, New Hampshire, Massachusett
and adjacent states, and given the name New England by Captai
John Smith (1580–1631), an English explorer, in the year 1614
World-wide distribution migrating to warmer seas in the winter.

Suborder ODONTOCETI
odous (Gr), genitive *odontos*, a tooth *ketos* (Gr) any sea monster,
whale.

Family HYPEROODONTIDAE (or ZIPHIIDAE) 18 specie
hyperōē (Ionic Gr) the upper part of the mouth, the palate *od*
(Ionic Gr), genitive *odontos*, a tooth; a misleading name, as the

usually have only two teeth, and these are in the lower jaw. However, on the palate there are small protuberances which were mistaken for teeth by the French zoologist Count de Lacépède (1756-1825) when he examined two of these whales stranded on the beach near Le Havre, France, in 1788. This whale inhabits the North Atlantic Ocean.

Bottle-nosed Whale *Hyperoodon ampullatus*
ampulla (L) a flask, a bottle *-atus* (L) suffix meaning provided with; a reference to the long snout in front of the bulging forehead. Widespread in the Northern Hemisphere, particularly the Arctic Ocean, and fairly common round British coasts.

Sowerby's Whale *Mesoplodon bidens*
mesos (Gr) middle, in the middle *oplon* (Gr) any tool or weapon *ōdon* (= *odous*) (Gr) a tooth *bidens* (L) having two teeth; there are usually only two teeth situated about the middle of the lower jaw. James Sowerby (1757-1822) was an English naturalist and artist; he was the first to describe this whale, in 1804. Sometimes known as the Beaked Whale or the Cowfish, it inhabits the North Atlantic and European waters.

True's Beaked Whale *M. mirus*
mirus (L) wonderful, extraordinary; this whale is remarkable for having the two teeth at the extreme tip of the lower jaw instead of in the middle, as *M. bidens* (above). F. W. True, an American zoologist, was an authority on whales; he was studying whales and writing about his findings during the years 1885 to 1913 and he described this whale in 1913. Less well known than most whales, it lives in the North Atlantic.

Stejneger's Beaked Whale *M. stejnegeri*
Dr L. Stejneger (1851-1943) was an American zoologist and author whose particular subject was taxonomy. This whale inhabits the northern Pacific Ocean.

Cuvier's Beaked Whale *Ziphius cavirostris*
ziphius (New L) derived from *xiphos* (Gr) a sword; this refers to the beak, or snout *cavus* (L) hollow *rostrum* (L) in animals, the snout; there is a hollow formation at the base of the snout. Baron Georges Cuvier (1769-1832) was the famous French comparative anatomist and Professor of Natural History. It has a world-wide distribution and is known round British coasts.

Family MONODONTIDAE 2 species
monos (Gr) single, alone *odous* (Gr), genitive *odontos*, a tooth.

White Whale or Beluga *Delphinapterus leucas*
delphis (also *delphin*) (Gr) the dolphin *a-* (Gr) prefix meaning not, or there is not *pteron* (Gr) a feather or wing; can mean a fin; this whale has no dorsal fin *leukos* (Gr) white; the adult whale is almost pure white *beluga* is said to derive from *belyi* (Russ) white. Inhabiting mostly Arctic seas.

Narwhal *Monodon monoceros*
Monodon, see above *keras* (Gr) the horn of an animal. They have a few rudimentary teeth when young but these never develop, except in the males which retain one canine; this grows to become an enormous horn, projecting forward to a length of over 2 m (7 to 8 ft) and twisted in an anticlockwise spiral. Narwhal comes from Old Norse, meaning 'corpse-whale'; their pallid colour is said to resemble that of a floating corpse. Inhabiting cold northern seas.

Family PHYSETERIDAE 3 species
phusētēr (Gr) a blow-pipe or tube; this refers to the blow-hole on the top of the head.

Sperm Whale or Cachalot *Physeter catodon* (or *macrocephalus*)
kata (Gr) down, below *odous* (Gr), genitive *odontos*, a tooth; there are up to 30 teeth on each side of the lower jaw, but few in the upper jaw and these are vestigial and non-functional *cachalot* (Fr) the sperm whale. Inhabiting all tropical and sub-tropical seas and occasionally colder areas.

Pygmy Sperm Whale *Kogia breviceps*
Kogia, a barbarous unmeaning name but it might be a tribute to a Turkish naturalist named Cogia Effendi who observed whales in the Mediterranean, in the early part of the nineteenth century *brevis* (L) short *ceps* (New L) derived from *caput* (L) the head; the head is short, and smaller than that of *Physeter catodon*, whose head is enormous. A rare species but probably world-wide distribution.

Dwarf Sperm Whale *K. simus*
simus (L) snub-nosed; it has a distinctly upturned snout. This rare species has been found in the Indian Ocean.

Family   PLATANISTIDAE   4 species

platus (Gr) flat, hence *platanos* (Gr) the plane-tree, so called from its broad flat leaf   *platanistēs* (Gr) or *platanista* (L) is a rare word, apparently only used by Pliny to mean some kind of fish in the Ganges. These dolphins have somewhat flattened beaks, used for digging in the mud of rivers.

Gangetic Dolphin or Susu   *Platanista gangetica*

(*Platanista* or *Susu*)

-*icus* (L) suffix meaning belonging to   *gangetica*, of the Ganges; it is also found in the Indus and other rivers of southern Asia. Susu is a Bengalese name for this dolphin.

La Plata Dolphin   *Pontoporia blainvillei* (*Pontoporia* or *Stenodelphis*)

pontos (Gr) the sea   *poreuō* (Gr) I carry (on land) or I ferry (on water) H. D. de Blainville (1777–1850) was a French Professor of zoology. This dolphin inhabits the Rio de la Plata in South America.

Amazonian Dolphin or Bouto   *Inia geoffrensis*

Inia is a Bolivian name for a dolphin   *geoffrensis* is after Etienne Geoffroy Saint-Hilaire (1772–1844), a French zoologist; the suffix *ensis* (L) meaning belonging to, is usually a reference to a locality rather than a person. Bouto is a Portuguese name for this dolphin; it is found only in the Amazon and other rivers in that part of South America.

Chinese River or White Flag Dolphin   *Lipotes vexillifer*

leipō (Gr) I am left behind, as in a race   -*tes* (Gr) suffix meaning pertaining to; referring to this dolphin as an isolated relict species; it was not named until 1918   *vexillum* (L) a standard, a flag   *fero* (L) I bear, I carry; this refers to the dorsal fin often seen above the surface of the water. Sometimes known as the Chinese Lake Dolphin, it is found in a lake several hundred kilometers up the Yangtze River in China, and also other rivers in South-East Asia.

Family   STENIDAE   8 species

Named in honour of Dr Nikolaus Steno (1638–1687), a celebrated Danish anatomist, geologist and author.

Guiana River Dolphin   *Sotalia guianensis*

Sotalia, a coined name of unknown origin   -*ensis* (L) suffix meaning belonging to; 'of Guiana'. Inhabiting the coast of north-eastern South America and some rivers in that area.

Plumbeous Dolphin   *S. plumbea*
plumbus (L) lead, so *plumbeus* (L) leaden, lead-coloured. Inhabiting the Indian Ocean.

Rough-toothed Dolphin   *Steno bredanensis*
Steno, see above under Family   *-ensis* (L) suffix meaning belonging to; named after a Monsieur Van Breda of Ghent, who sent a sketch of the skull of this dolphin to Georges Cuvier who attributed it to another species he had already named. However, it was a new species, and the present name was given by René Primevère Lesson (1794-1849) a French zoologist and surgeon, and best known as an ornithologist. He was the zoologist on the voyage of La Coquille during the years 1822 to 1825. The surfaces of the teeth of this dolphin are furrowed with ridges; it has a world-wide distribution.

Family   PHOCAENIDAE   6 species, possibly 7
phōkaina (Gr) a porpoise.

Common Porpoise   *Phocaena phocoena*
Porpoises, though similar to dolphins, have blunter snouts, lacking the beaks of dolphins. This porpoise is widespread in the waters of the Northern Hemisphere.

Burmeister's Porpoise   *P. spinipinnis*
spina (L) a thorn, can mean the spines of an animal, for example a porcupine   *pinna* (L) a feather, a wing, can mean a fin; 'spiky-finned' Dr K. H. K. Burmeister (1807-1892) was a zoologist and at one time Director of the Zoological Museum at Halle University in Germany. This porpoise inhabits both the west and east coasts of South America.

Black Finless Porpoise   *Neophocaena phocoenoides*
(*Neophocaena* formerly *Neomeris*)
neos (Gr) new   *phōkaina* (Gr) a porpoise   *-oides* (New L) suffix derived from eidos (Gr) shape, form; can mean a species, a sort; the name indicates a new subdivision of porpoises. There is no dorsal fin; inhabits the Pacific and Indian Oceans.

Dall's or White-flanked Porpoise   *Phocaenoides dalli*
Phocaenoides, see above   William H. Dall (1845-1927) was an Amer

can naturalist, and worked with the US Geographical Survey from 1884 to 1909. This porpoise inhabits the northern part of the Pacific Ocean.

Family DELPHINIDAE about 30 species
delphis, also *delphin* (Gr) the dolphin.

Blackfish or Pilot Whale *Globicephala melaena*
globus (L) a round ball, a globe *kephalē* (Gr) the head *melas*, also *melaina* (Gr) black, dusky; the head is short and the forehead bulging, and the colour is black except for the white throat. It would be more correctly called a dolphin. They travel in large schools, with one apparently leading, hence the name 'pilot'. Inhabiting the Atlantic and Pacific Oceans and also European waters, including the British Isles.

Indian Pilot Whale *G. macrorhyncha*
makros (Gr) long *rhunkhos* (Gr) beak, snout. It is found in the Pacific, Atlantic and Indian Oceans.

Irrawaddy River Porpoise *Orcaella brevirostris*
orca (L) a whale *-ellus* (L) diminutive suffix; 'a small whale'; it should be remembered that although some of the animals in this order are known as dolphins or porpoises, because of certain differences in their structure, strictly speaking they are all whales *brevis* (L) short *rostrum* (L) the beak or snout; the Irrawaddy is a river in Burma. Also found in the Indian and southern Pacific Oceans.

Killer Whale *Orcinus orca*
orca (L) a whale, which gave rise to orc, or ork, meaning a mysterious sea monster of horrible aspect *-inus* (L) suffix meaning like, pertaining to. It is known as the Killer Whale because it is an aggressive predator, killing and eating seals and porpoises as well as fishes. Sometimes known as the Grampus (see Risso's Dolphin, below) it has a world-wide distribution including the coast round the British Isles.

False Killer *Pseudorca crassidens*
seudēs (Gr) false *orca* (L) a whale, the killer whale; 'false killer' because in some respects it resembles the killer whale *crassus* (L) thick, heavy *dens* (L) a tooth; it has large powerful teeth, but circular in cross-section instead of oval as in the killer. This whale was first described from a fossilised skull by Sir Richard Owen in 1846, which

was found in 'the great fen of Lincolnshire . . . near Stamford'; it resembled that of the Killer Whale. Present day knowledge of the False Killer has been obtained almost entirely from examination of the bodies of these whales that have become stranded ashore, sometimes in large numbers. This is because although they have a worldwide distribution, they are seldom seen. Sir Richard Owen (1804–1892) was a distinguished anatomist and Director of the British Museum (Natural History) from 1856 to 1883.

Northern Right Whale Dolphin   *Lissodelphis borealis*
lissos (Gr) smooth   *delphis*, also *delphin* (Gr) the dolphin; a 'smooth-skinned' dolphin   *boreus* (L) northern   *-alis* (L) suffix meaning relating to   *borealis* (L) northern (a rare Latin word). It inhabits the northern Pacific Ocean.

Commerson's Dolphin   *Cephalorhynchus commersonii*
kephalē (Gr) the head   *rhunkhos* (Gr) the snout, beak; the whole head is curved, or beaked, not the snout only as in most dolphins. Dr P. Commerson (1727–1773) was a naturalist who worked with the French world navigator Vice-Admiral Baron de Bougainville (1729–1811) during the years 1766 to 1769. This dolphin inhabits the Pacific and Atlantic coasts of South America.

Common Dolphin   *Delphinus delphis*
delphis, also *delphin* (Gr) the dolphin   *-inus* (L) suffix meaning like. As the English name implies it is frequently seen at sea, often leaping clear of the water and playing round ships' bows. World-wide distribution, particularly warm and temperate seas, including British waters.

Pacific or Baird's Dolphin   *D. bairdi*
Professor S. F. Baird (1823–1887) was an American zoologist who did a survey of the Pacific Ocean during the years 1857 to 1859. This dolphin inhabits the southern Pacific in the Australia and New Zealand area.

Risso's Dolphin   *Grampus griseus*   (*Grampus* or *Grampidelphis*)
The origin of the word grampus is probably Spanish *grande pez*, 'great fish', and has come to mean 'one who puffs'   *griseus* (New L) grey from *gris* (Sp) grey; the general colour is grey, though lighter on the head, with possibly a white belly. A. Risso (1777–1845) was an Italian naturalist. World-wide distribution.

Bornean Dolphin *Lagenodelphis hosei*
lagēnos (Gr) a flagon *delphis* (Gr) the dolphin, 'bottle-shaped dolphin' Dr C. Hose (1863–1929) was a naturalist who lived in Sarawak from 1884 to 1907; little is known of this dolphin but one skeleton was found in Sarawak.

White-sided Dolphin *Lagenorhynchus acutus*
lagēnos (Gr) a flagon *rhunkhos* (Gr) snout, beak, 'flagon-nosed' *acutus* (L) pointed; it has a short distinct beak about 5 cm (2 in) long; there is a light broad band along the flanks. Inhabiting the northern part of the Atlantic.

White-beaked Dolphin *L. albirostris*
albus (L) white *rostrum* (L) the snout, beak. Inhabiting the northern part of the Atlantic including European waters and fairly common in the North Sea.

Bridled Dolphin *Stenella frontalis* (*Stenella* or *Prodelphinus*)
stenos (Gr) narrow *-ellus* (L) diminutive suffix *frons* (L), genitive *frontis*, the forehead *-alis* (L) suffix meaning pertaining to; a slender dolphin with a striped pattern on the head. This taxon has been called 'A genus in chaos' as the systematics are still very doubtful; about ten species have been described. This species inhabits the Atlantic and Indian Oceans.

Long-beaked Dolphin *S. longirostris*
longus (L) long *rostrum* (L) the beak, snout. Inhabiting the Pacific Ocean.

Bottle-nosed Dolphin *Tursiops truncatus* (*Tursiops* or *Tursio*)
tursio (L) a kind of fish resembling the dolphin; a name used by Pliny *opsis* (Gr) aspect, appearance *ops* (Gr) eye, face; 'looking like a dolphin' *trunco* (L) I shorten, I cut off, hence *truncatus* (New L) shortened; they have shorter beaks than other dolphins. This is the friendly dolphin, easily tamed and taught to perform in 'dolphinaria'; is widespread in the Atlantic, and found in the Bay of Biscay, the Mediterranean, and round the coasts of the British Isles.

Dogs, Weasels, Lions and their kin

CARNIVORA

In this order we shall find many familiar animals such as dogs, cats, weasels, badgers, otters, bears; and of course the 'big cats' such as the lion and tiger.

Although the name Carnivora means flesh-eating, some of these animals have a partly vegetarian diet and some are almost complete vegetarians. As will be seen, they differ enormously in outward form and size—for example, from a small weasel to an animal the size of a bear or a lion. However, a study of their anatomy shows them to be all related.

For purposes of classification they are divided into two suborders: Canoidea, the 'dog-like', and Feloidea, the 'cat-like'.

Subclass   EUTHERIA

Order   CARNIVORA

caro (L), genitive *carnis*, flesh   *voro* (L) I devour.

Suborder   CANOIDEA

canis (L) a dog   *-oides* (New L) from *eidos* (Gr) apparent shape, form; can mean a kind, sort. (Canoidea or Arctoidea.)

Family   CANIDAE   37 or more species

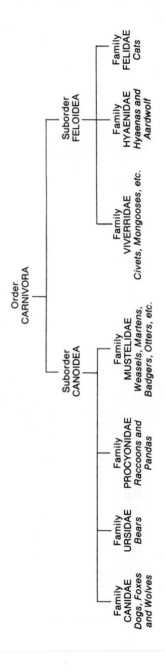

Grey Wolf *Canis lupus lupus*

canis (L) a dog *lupus* (L) a wolf. Inhabiting the wilder parts of northern Europe, North America and Canada, and Asia.

Red Wolf *C. l. niger*

niger (L) black; a misleading name; it was given to a melanistic individual, which by chance was black, or nearly black. (This phenomenon occasionally occurs, especially in mammals and birds, and is called melanism, from the Greek *melas*, black.) Normally, the coat is tawny. Inhabiting southern parts of North America, it is very rare and practically extinct.

Coyote *C. latrans*

latro (L) I bark, hence *latrans* (L) a barker; coyote is derived from *coyotl*, a Mexican name. Inhabiting North America.

Golden or Common Jackal *C. aureus*

aureus (L) golden. Inhabiting south-eastern Europe, Central and northern Africa, the Middle East and a large part of southern Asia, including Thailand but not Malaysia.

Black-backed Jackal *C. mesomelas*

mesos (Gr) middle *melas* (Gr) black. Living in Sudan and south to the Cape, on the eastern side of Africa.

Side-striped Jackal *C. adustus*

aduro (L) I set fire to, so *adustus* (L) burnt, sunburnt; it is a brownish colour with dark stripes on the sides. Inhabiting large areas of Africa, including Congo, Kenya and south to the Cape.

Simenian Jackal *C. simensis*

ensis (L) suffix meaning belonging to; it inhabits the Simien region and mountains in the northern part of Ethiopia and is one of the world's rarest animals.

Rüppell's Fox *C. rueppelli*

Dr W. P. E. S. Rüppell (1794–1884) was a German zoologist who travelled in North Africa during the years 1822 to 1827. This fox inhabits the Near and Middle East.

Dog *C. familiaris*

familiaris (L) domestic, belonging to a household, a servant; can mean familiar friend. World-wide.

European Red Fox *Vulpes vulpes vulpes*
vulpes (L) a fox. Inhabiting North America, Europe, the northern part of Africa, Asia and India. (See Tautonyms, page 13.)

North American Red Fox *V. v. fulva*
fulvus (L) tawny, yellowish-brown. A subspecies of *V. vulpes*.

Desert Fox *V. leucopus*
leukos (Gr) white *pous* (Gr) the foot. Inhabiting Iraq, Iran and the western part of India.

Arctic Fox *Alopex lagopus*
alōpēx (Gr) a fox *lagōs* (Gr) a hare *pous* (Gr) the foot; a hare's foot is hairy; like the polar bear, this fox has hair on the soles of its feet, probably to prevent slipping on the ice and as a protection against the cold. The range is widespread in the Arctic Circle.

Fennec Fox *Fennecus zerda*
Fenek is an Arabic word for a small fox *zerda* (Arab) derived from *zardawa*, a fennec. Inhabiting northern Africa.

Grey Fox *Urocyon cinereoargenteus*
oura (Gr) the tail *kuōn* (Gr) a dog; a reference to the tail because it has a concealed mane of stiff hairs, without any soft fur intermixed *cinereus* (L) ash coloured *argenteus* (L) silvery. Inhabiting southern Canada, the USA, Central America and the north-west corner of South America.

Island Grey or Beach Fox *U. littoralis*
litoralis (L) of the shore; although in English it is 'littoral', in Latin there should only be one 't'. However, as the name has long been internationally accepted the incorrect spelling must remain (see page 15). This fox lives on some islands off the coast of California.

Raccoon Dog *Nyctereutes procyonoides*
nux (Gr), genitive *nuktos*, night *ereuna* (Gr) an enquiry, a search hence *ereunētēs* (Gr) a searcher; 'a night-hunter' *pro* (Gr) before, in front of *kuōn* (Gr) a dog *-oides* (New L) from *eidos* (Gr) apparent shape, form; indicating its remarkable likeness to the raccoon *Procyon* (page 163). They are night-hunters and plant food also forms part of their diet. They live in eastern Asia, including Vietnam and Japan and have been introduced into Russia and Poland. They must not be confused with the common raccoon *Procyon lotor*, which lives in America.

Azara's or Western Dog   *Dusicyon gymnocercus*
dusis (Gr) a setting of the sun; *dusis helion*, 'setting sun'; can mean the west, western   *kuōn* (Gr) a dog   *gumnos* (Gr) naked   *kerkos* (Gr) the tail. Named after D. Felix de Azara (1746–1811), a zoologist who made expeditions to Paraguay during the years 1781 to 1801. This dog inhabits the forests of Paraguay.

Small-eared Fox or Zorro   *Atelocynus microtis*
atelēs (Gr) imperfect   *kuōn* (Gr) a dog; considered imperfect on account of the small ears   *mikros* (Gr) small   *ous* (Gr), genitive *ōtos*, the ear   *Zorra* is Spanish for a fox. A rare animal, inhabiting South America.

Crab-eating Fox   *Cerdocyon thous*
cerdō (Gr) the wily one, a fox   *kuōn* (Gr) a dog   *thōs* (Gr), genitive *thōos*, the jackal. Small mammals and birds are the main foods, and although it may occasionally eat crabs there is no reliable confirmation of this; the name dates from very early days of animal discovery. Inhabiting South America.

Maned Wolf   *Chrysocyon brachyurus*
chrusos (Gr) golden   *kuōn* (Gr) a dog; it has a handsome coat of long reddish-brown hair   *brakhus* (Gr) short   *oura* (Gr) the tail; it has a mane on the neck and shoulders which is erected when the animal is excited. Inhabiting South America.

Bush Dog   *Speothos venaticus*
speos (Gr) a cave   *thōs* (Gr) the jackal; they dig and live in holes in the ground   *venaticus* (L) for the chase, belonging to hunting. Inhabiting the Amazonian forest area.

Dhole   *Cuon alpinus*
kuōn (Gr) a dog   *alpinus* (L) alpine; 'alpine' probably refers to the slopes at the foot of mountains because they live on the plains as well as in the hills. The anatomy of the dhole, for example 40 teeth instead of the usual 42, and the convex profile of the skull, separate it somewhat from typical Canidae. Dhole is an East Indian name. It inhabits India but not Sri Lanka and is widespread in south-eastern Asia, including parts of Russia, China and Korea.

Cape Hunting Dog   *Lycaon pictus*
lukos (Gr) a wolf, so *lukaon* (Gr) a wolf-like animal   *pingo* (L) I paint, hence *pictus* (L) painted; the coat looks as though it had been care-

lessly daubed with splodges of white and orange on a dark background. Inhabiting the southern half of Africa.

Bat-eared Fox *Otocyon megalotis*
ous (Gr), genitive *ōtos*, the ear *kuōn* (Gr) a dog; an allusion to the large ears *megas* (Gr) big, wide. Inhabiting Africa.

Family URSIDAE 8 species
ursus (L) a bear.

Spectacled Bear *Tremarctos ornatus*
trema (Gr) a hole *arktos* (Gr) a bear; a reference to an unusual hole in the humerus bone *ornatus* (L) dress, equipment; it has yellow rings round the eyes giving the appearance of spectacles. Inhabiting the Andes Mountains of South America.

Moon Bear or Himalayan Black Bear *Selenarctos thibetanus*
selēnē (Gr) the moon *arktos* (Gr) a bear; it has a white mark on the chest like a crescent moon *thibetanus*, of Tibet.

Brown Bear *Ursus arctos*
ursus (L) a bear *arktos* (Gr) a bear. It varies in colour from very pale brown to black. Inhabiting North America from Alaska south to northern Mexico, central Europe where there are forests and mountains and most of Eurasia.

American Black Bear *U. americanus*
Occasionally brown, grey or even white, in colour, their range widespread in North America.

Polar Bear *Thalarctos maritimus*
thalassa (Gr) the sea *arktos* (Gr) a bear *maritimus* (L) of the sea Living among the Arctic pack-ice.

Malayan or Sun Bear *Helarctos malayanus*
helē (= *eilē*) (Gr) the heat of the sun *arktos* (Gr) a bear; although according to Palmer 'Probably referring to its tropical habitat', it more likely an allusion to a yellow mark on the chest said to represent the sun *-anus* (L) suffix meaning belonging to; it is not confined Malaya and is widespread in that area.

Sloth Bear *Melursus ursinus*
mel (L) honey *ursus* (L) a bear *-inus* (L) suffix meaning like, belong

ing to; they are fond of honey and tear open and eat ants' and bees' nests. Known as 'sloth bears' because they usually move with a slow, shambling gait. Inhabiting Sri Lanka and India.

Family **PROCYONIDAE** 18 species

pro- (Gr) prefix meaning before, in front of *kuōn* (Gr) a dog; this is the raccoon family; it is considered that the ancestors of the dogs were probably raccoons, hence 'before dogs'.

Cacomistle or Cacomixtle *Bassariscus astutus*

bassara (Gr) a fox; of Thracian origin with the same meaning as *alopex* (Gr) a fox *-iscus* (L) diminutive suffix *astutus* (L) cunning. Cacomixtle is a Mexican name for this animal; sometimes known as the Ringtailed Cat, it inhabits Mexico and ranges north to south-western USA as far as southern Oregon and Alabama.

Guayonoche or American Cacomistle *B. sumichrasti*

Professor F. E. Sumichrast (1828–1882) was a Mexican zoologist. Guayonoche is a Spanish name for this small fox-like animal, which inhabits the forest areas of southern Mexico and Central America.

North American Raccoon *Procyon lotor*

Procyon, see Family above *lotor* (New L) a washer, from *lavo* (L) I wash; when in captivity, this raccoon has been seen to wash its food before eating, but there are no reliable reports of this behaviour concerning raccoons in the wild. The reason for the behaviour seen in captive animals is not really established but is probably not for cleansing purposes. This raccoon inhabits North and Central America.

Crab-eating Raccoon *P. cancrivorus*

cancer (L) a crab *voro* (L) I devour; raccoons eat a wide variety of small animals and some plants, and the animal part of the diet includes shellfish. Inhabiting river valleys and coastal areas from Panama to the south of the Amazon.

Red or Ringtailed Coatimundi *Nasua nasua*

nasus (L) the nose; the nose is exceptionally long, flexible and almost like a trunk. Coatimundi is a Tupi name for this raccoon-like animal which inhabits the forests of Central and South America.

White-nosed Coati *N. narica*

naris (L) the nostril *-icus* (L) suffix meaning belonging to; it has a long nose like that of *N. nasua*, above. A small raccoon-like animal living in southern Mexico and ranging south to Panama.

Nelson's Coati   *N. nelsoni*
Named in honour of Dr Edward Nelson (1855-1934), an American zoologist and explorer. This coati inhabits an island off the coast of Quintana Roo, Mexico.

Kinkajou   *Potos flavus*
Potos is derived from *Potto*, the native name for this animal   *flavus* (L) yellow; it has gingery-yellow fur   Kinkajour is a South American Indian name. It inhabits the forests of Central and South America. It is interesting to note that another animal, living in Africa, and known locally by the name Potto, is rather similar in appearance but completely unrelated; it is a primate (page 83).

Bushy-tailed Olingo   *Bassaricyon gabbii*
bassara (Gr) a fox; of Thracian origin with the same meaning as *alopex* (Gr) a fox   *kuōn* (Gr) a dog   Professor W. M. Gabb (1839-1878) was an American scientist who was in Costa Rica in 1876. This small raccoon-like animal, similar to the kinkajou, lives in Central and South America.

Beddard's Olingo   *B. beddardi*
Dr F. E. Beddard, FRS (1858-1925) was an English zoologist and at one time President of the Zoological Society of London. This olingo known locally as the Cuataquil, inhabits South America; it may be the same animal as *Bassaricyon gabbii* (above).

Cat-Bear or Red Panda   *Ailurus fulgens*
aiolos (Gr) quick moving, wriggling   *oura* (Gr) the tail, hence *ailouro* (Gr) a cat; Liddell and Scott's Greek Lexicon says 'as expressive, no of colour, but of the *wavy motion of the tail* peculiar to the cat kind' a 'tail-waver'   *fulgens* (L) shining, gleaming; it has a brightly coloured shiny coat   *panda* is originally an East Indian word. Sometimes known as the Lesser Panda, it inhabits Nepal, the Himalayas, northern Burma, Szechwan and Yunnan.

Giant Panda   *Ailuropoda melanoleuca*
ailouros (Gr) a cat (see Cat-Bear above)   *pous* (Gr), genitive *podos* a foot; a reference to the likeness of the feet to those of *Ailurus*; *mela* (Gr) black   *leukos* (Gr) white. It is strange that this carnivore has reputation for eating only bamboo shoots; in fact it will also eat sma

animals such as rodents, birds and fishes. It inhabits the bamboo forests of Eastern Tibet and Szechwan in south-western China.

Family MUSTELIDAE 70 species
mustela (L) a weasel.

Stoat or Short-tailed Weasel *Mustela erminea erminea*
ermine (Old Fr) *hermine* (Fr) a stoat. Normally the coat is brown with white beneath; it may change to white in winter, when it is used for robes and known as ermine. Widespread in northern Europe and northern North America, including Canada and the Arctic.

Irish Stoat *M. e. hibernica*
Hibernia (L) Ireland *-icus* (L) suffix meaning belonging to.

Islay Stoat *M. e. ricinae*
ricinium (L) a small veil. Islay is a small island of the Inner Hebrides, Scotland, and the Romans gave it the name Ricina because it was always veiled in mist.

Weasel *M. nivalis nivalis*
nix (L), genitive *nivis*, snow, so *nivalis* (L) of snow, snowy; the weasel is a reddish-brown colour and white beneath, though a northern form turns quite white in winter. Inhabiting North America, Europe, North Africa and Asia.

Least Weasel *M. n. rixosa*
rixa (L) a fight, a quarrel *-osus* (L) suffix meaning full of, prone to; 'quarrelsome'. Inhabiting North America.

Black-footed Ferret *M. nigripes*
niger (L) black *pes* (L) a foot. A weasel very similar to the domesticated ferret of Britain and Europe; it inhabits North America.

American Mink *M. vison*
vison (Fr) the American mink. Vison is also used as an English name for this animal.

European Mink *M. lutreola*
lutra (L) the otter *-olus* (L) diminutive suffix, a small otter; it swims and dives well. Probably widespread in Russia but becoming scarce or even unknown in certain parts of Europe.

Asiatic or Russian Polecat *M. eversmanni*
Professor Dr E. Eversmann (1794–1860) was a German zoologist.

European Polecat   *M. putorius putorius*
putor (L), genitive *putoris*, a bad smell; polecats can make a foul-smelling discharge from glands under the tail.

Common Ferret   *M. p. furo*
fur (L) a thief, from which the English name 'ferret' is derived. This is the ferret that is used for hunting. It is probably a domesticated descendant of the Russian or Steppe polecat, possibly *M. eversmanni*. It is an albino, having an almost white coat and pink eyes, and is unknown except in captivity.

Stripe-bellied Weasel   *Grammogale africana*
gramme (Gr) a mark, a line   *galē* (Gr) a weasel or marten-cat; it has a chestnut-brown stripe along the belly   *africana* is misleading; it was first described as an African weasel by Professor A. G. Desmarest in 1818, but this was incorrect as it had not been collected in Africa. It was not until 1913 that Dr Angel Cabrera y Latorre (1879–1960), a Spanish zoologist, established that it was a Brazilian weasel, and even then the amendment was neglected by other mammalogists until 1937. However, by the rules of the International Commission on Zoological Nomenclature the specific name must remain *africana*; 'The first scientific name given to an animal after Jan 1st 1758 stands, even though it is not descriptively accurate.' This weasel inhabits the Amazon basin in Brasil.

Mottled or Marbled Polecat   *Vormela peregusna*
T. S. Palmer, in his standard work *Index Generum Mammalium* (1904) gives *Vormela* as Latin derived from the German, and quotes '*Anima cujus Agricola sub nomine* Vormelae *(Germanice Wormlein) mentionem fecit*. (Pallas). This could be translated as 'An animal whom Agricola cited under the name of Vormela.' *peregusna* is a Latinised form of *pereguznya*, the Ukrainian name for the polecat, after Professor S. I Ognev (1866–1951), a Professor at Moscow University. The Mottled Polecat inhabits south-eastern Europe and south-west and central Asia as far eastwards as the Gobi Desert.

Pine Marten   *Martes martes*
martes (L) a marten. Inhabiting Europe and central Asia.

Beech or Stone Marten   *M. foina*
foina (It dialect) a polecat. Inhabiting Europe, the USSR and the Himalayas.

Yellow-throated Marten   *M. flavigula*
flavus (L) yellow   *gula* (L) the throat. Inhabiting eastern Asia.

American Marten   *M. americana*
Inhabiting North America including Canada and Alaska.

Fisher or Pennant's Marten   *M. pennanti*
This refers to Dr T. Pennant, FRS (1726–1798), an English zoologist.
This marten inhabits North America. The name 'fisher' could be
misleading as it is not established that it catches live fish; it lives near
water but is not aquatic. Exceedingly rare, and a protected species.

Tayra   *Eira barbara*
Eira is the Guarani name for the Tayra; the Guaranis are a people of
Bolivia and Paraguay   *barbaros* (Gr) strange, foreign; originally
meaning all those who were not Greek. The Tayra is a giant weasel
of South America.

Grison   *Galictis vittata*
galē (Gr) a marten-cat or polecat   *iktis* (Gr) a marten-cat   *vittatus*
(L) bound with a ribbon; can mean striped   *grison*, from *gris* (Sp)
grey; the fur is blackish below and grey above. Inhabiting Central
and South America.

Patagonian Weasel   *Lyncodon patagonicus*
lunx (Gr) a lynx   *odon* (= *odous*) (Gr) a tooth; a reference to the
peculiarity of having only three pairs of molar teeth in each jaw.
Like the lynx it has only three cheek teeth behind the canines in both
upper and lower jaws, and so, like the lynx, has only 28 teeth; this is
a small number for a member of the Carnivora   *-icus* (L) suffix meaning belonging to, 'of Patagonia'. An uncommon species inhabiting
South America.

Striped Polecat   *Ictonyx striatus*
iktis (Gr) a marten-cat or polecat   *onux* (Gr) a claw; an allusion to
the strong non-retractile claws on the fore feet   *striatus* (L) striped;
it has a black coat with broad white stripes along the back. It is sometimes known as the Zorille or Zorilla, as *zorillo* is Argentine Spanish
for the skunk, a related animal. The American Striped Skunk *Mephitis
mephitis* is similar in appearance and behaviour. The Striped Polecat
is widespread in Africa.

Libyan Striped Weasel   *Poecilictis libyca*

poikilos (Gr) many-coloured, dappled   *iktis* (Gr) a marten-cat or polecat   *libyca*, of Libya. Inhabiting northern Africa.

White-naped Weasel   *Poecilogale albinucha*
poikilos (above)   *galē* (Gr) a marten-cat or weasel   *albus* (L) white *nucha* (New L) the neck. Inhabiting central and southern Africa.

Wolverine or Glutton   *Gulo gulo*
gula (L) the throat, so *gulosus* (L) gluttonous; it probably eats no more than other carnivores, though it has a reputation for greed. Living in North America, northern Europe, the Arctic and northern Asia.

Ratel or Honey Badger   *Mellivora capensis*
mel (L), genitive *mellis*, honey   *voro* (L) I devour; a reference to its favourite food   *capensis*, of Cape Province, South Africa. The name Ratel was originally South African Dutch, possibly from *raat* (Du) a honeycomb. Inhabiting Africa and southern Asia.

Eurasian Badger   *Meles meles*
meles (L) a badger. Inhabiting Europe and Asia; it ranges from the British Isles to China.

Hog Badger or Sand Badger   *Arctonyx collaris*
arktos (Gr) a bear   *onux* (Gr) a claw; a reference to the long slightly curved blunt claws, as in the bear   *collum* (L) the neck   *-aris* (L) suffix meaning pertaining to; there is a white patch on the neck Inhabiting southern Asia and Sumatra.

Teledu or Stink Badger   *Mydaus javanensis*
mudaō (Gr) I am wet, damp, from *mudos* (Gr) damp, decay; an allusion to the fetid skunk-like odour of the animal; like the skunk, it has a gland that secretes a foul-smelling liquid which it can expel a short distance   *-ensis* (L) suffix meaning belonging to; 'of Java'; it is also found in Sumatra and Borneo. Teledu is from the Malayan native name.

American Badger   *Taxidea taxus*
taxus (New L) a badger   *idea* (Gr) appearance; can mean kind, sort Inhabiting North America.

Chinese Ferret Badger   *Melogale moschata*
meles (L) a badger   *galē* (Gr) a marten-cat or polecat   *moskhos* (Gr) musk, hence *moschatus* (New L) musky; it has a rather strong smell Inhabiting southern Asia.

Javan Ferret Badger   *M. orientalis*
-*alis* (L) suffix meaning relating to, thus *orientalis*, from the Orient.
It lives in Java and Borneo.

Striped Skunk   *Mephitis mephitis*
mephitis (L) a noxious smell; skunks actually squirt a foul-smelling
liquid at an enemy with considerable accuracy. Inhabiting Hudson
Bay in the north and southwards through to Mexico.

Hooded Skunk   *M. macroura*
makros (Gr) large, long   *oura* (Gr) the tail. Living in Mexico and
Central America.

Spotted Skunk   *Spilogale putorius*
spilos (Gr) a stain, a spot   *galē* (Gr) a marten-cat or weasel   *putor*
(L), genitive *putoris*, a bad smell. The range is from the southern part
of North America and south to Panama.

Hog-nosed Skunk   *Conepatus leuconotus*
konis (Gr) dust   *pateō* (Gr) I walk; can mean to frequent a place;
it lives in open country and deserts and it may also refer to its burrowing habits   *leukos* (Gr) white   *notos* (Gr) the back. Inhabiting North
America and Central America.

Eurasian Otter   *Lutra lutra*
lutra (L) the otter. Inhabiting Europe including the British Isles,
North Africa, Asia, Java and Sumatra.

Canadian Otter   *L. canadensis*
ensis (L) suffix meaning belonging to; it inhabits North America
including Canada.

Indian Small-clawed Otter   *Amblonyx cinerea*
(or subgenus of *Aonyx*)
amblus (Gr) blunt, point taken off   *onux* (Gr) a claw; it has short
blunt claws   *cinis* (L), genitive *cineris*, ashes, hence *cineria* ash-
coloured. Inhabiting southern Asia, Borneo, Java, Sumatra and the
Philippines.

Cape Clawless Otter   *Aonyx capensis*
a- (Gr) prefix meaning not, or there is not   *onux* (Gr) a claw; the
claws are small and rudimentary   *capensis*, of Cape Province, South
Africa. It ranges north to the border of the Sahara.

Giant Otter or Saro   *Pteronura brasiliensis*
pteron (Gr) feathers; can mean wings   *oura* (Gr) the tail; this otter's
tail has lateral flanges that make it appear something like a wing;
'the fin-like dilatation on each side of the hinder half of the tail'
(Gray 1837)   *-ensis* (L) suffix meaning belonging to; also inhabiting
Guyana and Surinam and not confined to Brazil; very rare.

Sea Otter   *Enhydra lutris*
enhudris (Gr) an otter, from *enhudro-bios*, living in the water   *lutra* (L)
an Otter; this otter lives almost entirely in the water and rarely comes
ashore. Inhabiting the shore areas of the northern Pacific, it is a rare
animal and is protected by the governments of both the United States
and Russia.

Suborder   FELOIDEA (or AELUROIDEA)
see *Ailurus fulgens* (page 164)
feles (L), genitive *felis*, a cat   *-oides* (New L) from *eidos* (Gr) apparent
shape, form; can mean a kind, sort; 'cat-like'.

Family   VIVERRIDAE   about 82 species
viverra (L) a ferret; this is misleading as the animals in this family are
not ferrets, they are more 'cat-like'.

African Linsang   *Poiana richardsoni*
Poiana, apparently from 'Po' of Fernando Po, the island off the coas
of Cameroun, West Africa, where the species was found and described
-anus (L) belonging to; named after Dr John Richardson, Inspecto
of the Naval Hospital at Haslar, by T. R. H. Thompson, RN, surgeon
of 'the late Africa expedition' (1842).

Small Spotted or Feline Genet   *Genetta genetta*
genette (Fr) a genet or civet-cat. Inhabiting Africa and Spain.

Abyssinian Genet   *G. abyssinica*
-icus (L) suffix meaning belonging to; of Abyssinia; it also inhabit
other parts of Africa. It can be tamed and domesticated and wa
probably the original 'cat' of ancient Egypt.

Blotched Genet or Tigrine Genet   *G. tigrina*
tigris (L) a tiger, so *tigrinus* tiger-like. Inhabiting Africa.

Rusty-spotted Genet   *G. rubiginosa*
rubor (L) redness, hence *rubiginosus* (L) rusty. From South Africa.

Giant or Victorian Genet   *G. victoriae*
This genet takes its name from Lake Victoria, Africa. It lives in a belt
of country that surrounds this lake and is also found in the Ituri
Forest, Zaire.

Small Indian Civet   *Viverricula indica*
viverra (L) a ferret (see under Family above)   *-culus* (L) diminutive
suffix   *-icus* (L) suffix meaning belonging to; of India. It ranges from
India through Sumatra and Java to Bali.

Water Civet   *Osbornictis piscivora*
This civet, discovered as recently as 1916 at Niapu, in northern Zaire,
is named after Dr Henry F. Osborne (1857–1935), an American
zoologist; no live specimens have been found since then   *iktis* (Gr)
a marten-cat   *piscis* (L) a fish   *voro* (L) I devour. The extent of its
range is not established.

African Civet   *Viverra civetta*
viverra (L) a ferret (see above under Family)   *civette* (Fr) the civet-cat;
civet is the name of a highly valued perfume fixative obtained from
this animal. Widespread in Africa.

Large Indian Civet   *V. zibetha*
zibetto (It) a civet-cat. Inhabiting southern Asia.

Banded Linsang   *Prionodon linsang*
prión (Gr) a saw   *odón* (Gr) a tooth; a reference to the simple and
similar teeth which form a saw-like row   *linsang* is a Javanese name
for this animal which lives in Java, Sumatra, Borneo and southern
Asia.

Spotted Linsang   *P. pardicolor*
pardus (L) a leopard   *color* (L) colour; a reference to the leopard-like
spots. Inhabiting Nepal, Assam and Vietnam.

West African or Two-spotted Palm Civet   *Nandinia binotata*
Nandine is a West African native name for the palm civet   *bi-* (L) two
notatus (L) marked; it has two white spots on the shoulders.

Small-toothed Palm Civet   *Arctogalidea trivirgata*
arktos (Gr) a bear   *galé* (Gr) a marten-cat or weasel   *galideus* (Gr)
diminutive of *galé*, a young weasel; head conical, nose compressed,
broad forehead; seemingly bear-like   *tri* (L) three   *virgatus* (L)
made of twigs; can mean striped; it has three dark stripes along the

back. The first of these civets discovered were frequently seen in palm trees and so became known as Palm Civets. They inhabit large areas of India and China, and Sumatra, Java and Borneo.

Common Palm Civet or Musang   *Paradoxurus hermaphroditus*
para (Gr) beside; can mean contrary to, against   *doxa* (Gr) opinion; *para doxan* (Gr) contrary to opinion   *oura* (Gr) the tail; T. S. Palmer, in his standard work *Index Generum Mammalium* (1904), gives this explanation 'from the mistaken idea that the tail was prehensile. Though the tail is not prehensile the animal has the power of coiling it to some extent, and according to Blanford "in caged specimens the coiled condition not infrequently becomes confirmed and permanent".' Hermaphroditos, in the Greek legend, was the son of Hermes and Aphrodite; a nymph from the fountain at Salmacis was so attracted by him that she prayed for perpetual union with him, and the resulting form had the characteristics of both sexes. This civet is not a hermaphrodite (though some animals are, e.g. the earthworm *Lumbricus terrestris*), but probably it was found difficult to distinguish one sex from another, which is a feature common to many carnivores. It inhabits southern Asia and the islands of the Philippines, Malaysia and Indonesia.

Masked Palm Civet   *Paguma larvata*
Paguma is a word coined evidently from puma; the name was given by Dr J. E. Gray in 1831 and he gave a number of other obscure names of animals   *larvatus* (L) bewitched; can mean masked; the body is a uniform grey-brown but the black face has a conspicuous white stripe down the middle, hence 'masked'. It lives in the forest of south-east Asia, Sumatra and Borneo.

Binturong   *Arctictis binturong*
arktos (Gr) a bear   *iktis* (Gr) a marten-cat; it is rather bear-like and sometimes known locally as the Bear-Cat   *binturong* is the Malaya native name. Inhabiting southern Asia including the Himalaya, Burma, Malaya, Java and Sumatra.

Fanaloka or Malagasy Civet   *Fossa fossa*
Fossa or *foussa* is a native name for this civet, and not to be confused with the Fossa (see page 175), a different animal although they both live in Madagascar. Fanaloka is another native name.

Banded Palm Civet   *Hemigalus derbyanus*
hēmi- (Gr) half; a prefix, in this case used to indicate like, or similar

galē (Gr) a marten-cat or weasel; named after the Thirteenth Earl of Derby (formerly the Honourable E. S. Stanley) (1775-1851); he was President of the Zoological Society of London in 1831. This civet inhabits southern Asia, Sumatra and Borneo.

Owston's Banded Civet   *Chrotogale owstoni*

khrōs (Gr), genitive *khrōtos*, the skin; can mean colour of the skin, complexion   *galē* (Gr) a marten-cat or weasel; it has dark bands round the body. Alan Owston (1853-1915) was a zoologist who lived in Japan for many years. This civet inhabits Tonkin and Laos.

Otter Civet   *Cynogale bennetti*

kuōn (Gr), genitive *kunos*, a dog   *galē* (Gr) a marten-cat or weasel; it has an elongate face, much produced and compressed. Another rather obscure name by Gray (*see Paguma larvata*, page 172); it has an otter-like body, and is similar to an otter in its habits and diet. Gray gave the specific name for his friend Edward T. Bennett (1797-1836), zoologist and surgeon. He was prominent in starting the London Zoological Society and was the Secretary from 1831 until his death in 1836. He would have described the Otter Civet himself had he lived. It inhabits southern Asia, Sumatra and Borneo.

Falanouc   *Eupleres goudoti*

eu- (Gr) prefix meaning well   *pleres* (Gr) full; can mean complete; in reference to the full number of five toes on both fore and hind feet . Goudot, a French naturalist, found this animal in Madagascar about 1830. Falanouc or *falanaka* is a Malagasy native name. Sometimes known as the Small-toothed Mungoose (sic), the teeth are so small that it was originally believed to be an insectivore. Inhabiting Madagascar.

Giant Falanouc   *E. major*

major (= *maior*) (L) greater, larger. Inhabiting Madagascar.

Ring-tailed Mongoose   *Galidia elegans*

galē (Gr) a marten-cat or weasel   *galideus* (Gr) diminutive of *galē*, a young weasel or kitten   *elegans* (L) neat, elegant. Living in Madagascar.

Broad-striped Mongoose   *Galidictis striata*

galidia, see above   *iktis* (Gr) a marten-cat   *striatus* (L) striped. Inhabiting Madagascar.

Suricate or Slender-tailed Meerkat   *Suricata suricatta*

Surikate or *suricat* are South African native names   *meerkat* is South African Dutch, meaning 'lake cat'; although living in dry places in South Africa, it is fond of water.

Egyptian Mongoose   *Herpestes ichneumon*
herpestes (Gr) a creeper; an allusion to its stalking habits   *ikhneumon* (Gr) the tracker, an Egyptian animal of the weasel kind, or mongoose. Sometimes known as Pharaoh's Rat, it was highly regarded in ancient times because it hunted for crocodile eggs and ate them. Inhabiting Africa and southern Europe.

Small Indian Mongoose   *H. auropunctatus*
aurum (L) gold, golden   *pungo* (L) I puncture, thus *punctatus*, spotted as with punctures. Inhabiting southern Asia.

Crab-eating Mongoose   *H. urva*
Urva is a Nepalese name for this mongoose from southern Asia.

Dwarf Mongoose   *Helogale parvula*
helos (Gr) low ground by rivers, a marsh   *galē* (Gr) a marten-cat or weasel; a reference to its habitat   *parvus* (L) small, so *parvulus* (L) very small. Inhabiting Africa.

Marsh Mongoose   *Atilax paludinosus*
a- (Gr) prefix meaning not, or there is not   *thulax* (Gr) a pouch Cuvier says 'par la considération de toute absence de poche à l'anus' This refers to the alleged absence of the anal scent gland, but some say Cuvier was wrong when he named it as it does have the gland *palus* (L), genitive *paludis*, a marsh, a swamp   *-osus* (L) suffix meaning full of; i.e. 'marshy'. A good swimmer inhabiting central and southern Africa.

Banded Mongoose   *Mungos mungo*
Mungos is derived from *mangus* (Marathi); the Maratha are a people of the south central part of India. Although this mongoose inhabits East Africa the name derives from the Indian mongoose. It has about twelve dark transverse bands across the back.

Angolan Cusimanse (or Kusimanse)   *Crossarchus ansorgei*
krossoi (Gr) a fringe, tassels   *arkhos* (Gr) the rump, the hind part it has a bushy tail   Dr W. J. Ansorge (1850-1913), a naturalist and author, was in Angola in 1905. This mongoose lives in Angola and other parts of Central Africa. Cusimanse is from *kusimanse*, a native name, probably from Liberia.

White-tailed Mongoose   *Ichneumia albicauda*
ikhneumon (Gr) the tracker (see *Herpestes ichneumon*, page 000)   *albus*
(L) white   *cauda* (L) the tail of an animal. Inhabiting southern
Arabia and Africa.

Bushy-tailed Mongoose   *Bdeogale crassicauda*
bdeō (Gr) I break wind, I stink   *galē* (Gr) a marten-cat or weasel;
it can produce a foul-smelling glandular secretion similar to a skunk
crassus (L) thick, dense   *cauda* (L) the tail of an animal. Inhabiting
East Africa.

Meller's Mongoose   *Rhynchogale melleri*
rhunkhos (Gr) the snout, beak   *galē* (Gr) a marten-cat or weasel; a
reference to the pointed nose   Dr C. J. Meller (1836–1869) was
Superintendent of the Botanical Gardens in Mauritius in 1865; this
island in the Indian Ocean lies to the east of Madagascar. The mongoose, a very rare species, inhabits East Africa.

Yellow Mongoose   *Cynictis penicillata*
uōn (Gr), genitive *kunos*, a dog   *iktis* (Gr) a marten-cat; this means
intermediate between, or connecting, the dogs and civets   *penicillus*
(L) a painter's brush; it has a bushy tail with a white tip. Inhabiting
South Africa.

Fossa or Foussa   *Cryptoprocta ferox*
kruptos (Gr) hidden   *prōktos* (Gr) hinder parts, tail; 'hidden in the
hinder parts'; it has an anal gland that produces an evil-smelling
liquid, like the skunk   *ferox* (L) brave, fierce; it has a reputation for
being unusually fierce but this is now considered to be exaggerated.
Fossa is a Malagasy name, and the animal must not be confused with
the Malagasy Civet which unfortunately has been given the Latin
name *Fossa fossa* (see page 172); quite a different animal but they both
live in Madagascar.

Family   HYAENIDAE   4 species
huaina (Gr) the hyaena, from *hus* (Gr) a hog, on account of the bristly
mane.

Aardwolf   *Proteles cristatus*
prōtos (Gr) first; can mean in front   *teleos* (Gr) complete, perfect;
'complete in front'; referring to the aardwolf having five toes on the
fore feet but only four on the hind feet   *cristatus* (L) crested; all

hyaenas have a crest or mane that normally lies flat. However, some say the aardwolf's mane is permanently erected, but there is disagreement about this among zoologists. Aardwolf is a Dutch name meaning 'earth-wolf'. Inhabiting southern and eastern Africa.

Spotted Hyaena   *Crocuta crocuta*
crocus (L) the crocus; also the colour of saffron, yellow   *utus* (L) suffix meaning provided with   *crocuta* is a rare Latin word originally meaning 'an unknown wild animal of Ethiopia, perhaps the hyaena'; it has no other meaning, but there is a Greek word *krokōtos* meaning saffron-coloured. This hyaena is tawny to yellow with dark spots; the range is widespread in Africa south of the Sahara.

Brown Hyaena   *Hyaena brunnea*
Hyaena, see above under Family   *brunneus* (New L) dark brown. A rare hyaena in danger of extinction, inhabiting South Africa.

Striped Hyaena   *H. hyaena*
It has six vertical black stripes on the flanks. Inhabiting Africa, the Near East and India.

Family   FELIDAE   about 36 species
feles (L), genitive *felis*, a cat.

European Wild Cat   *Felis silvestris*
silva (L) a wood, hence *silvestris* (L) belonging to woods. Inhabiting western Asia, the wild parts of Europe, and still found in Scotland.

African Wild Cat or Mu   *F. libyca*
libyca, belonging to Lybia. The Egyptians called this cat the *mu*, and trained it to hunt mice; it is not confined to northern Africa and also inhabits Asia.

Cat   *F. catus*
catta (New L) a cat; the domestic cat, common throughout the world and sometimes known as *F. domestica*.

Serval   *F. serval*
serval (Port) a deer-wolf, transferred from another animal. Inhabiting Africa.

Sand Cat   *F. margarita*
Named after a Général Margueritte on duty in Algeria in the 1850's. From North Africa and south-west Asia.

Jungle Cat   *F. chaus*
Chaus is an ancient name for a wildcat of Africa but the origin of the word is obscure. Inhabiting North Africa and Asia.

Leopard Cat   *F. bengalensis*
-ensis (L) suffix meaning belonging to; it is not confined to Bengal and is widespread in south-eastern Asia.

Black-footed Cat   *F. nigripes*
niger (L) black   *pes* (L) a foot. Inhabiting South Africa.

Persian Lynx or Caracal   *F. caracal*
caracal is from *karakulak* (Turk) meaning black ear; it has a fawn coat and large black ears with black tufts. It inhabits Africa, Arabia and India.

Northern Lynx   *F. lynx lynx*
lynx (L) a lynx   *lunx* (Gr) a lynx; from its light colouring and bright eyes and akin to Old English *lox*, and Old High German *luhs*, and Greek *leukos*, meaning white, bright. Inhabiting northern Asia, Europe and North America.

Canada Lynx   *F. l. canadensis*
ensis (L) suffix meaning belonging to; 'of Canada'.

Bobcat   *F. rufa*
rufus (L) red, ruddy. Widespread in the USA and Mexico.

Pallas's Cat   *F. manul*
manul (Mongolian) a small wildcat. Professor P. S. Pallas (1741–1811) was a German zoologist and explorer. This wild cat lives in the mountains of Mongolia, Siberia and Tibet.

Marbled Cat   *F. marmorata*
marmor (L) marble, hence *marmoratus* (L) marbled. Inhabiting southern Asia, Sumatra and Borneo.

African Golden Cat   *F. aurata*
aurum (L) gold, the colour of gold, so *aureatus* (L) adorned with gold. Inhabiting Africa.

Fishing Cat   *F. viverrina*
viverra (L) a ferret   *-inus* (L) suffix meaning like, pertaining to. Inhabiting southern Asia, Sumatra and Java.

Ocelot   *F. pardalis*
pardus (L) a panther, a leopard (see Leopard, below)   *-alis* (L)
suffix meaning relating to. Ocelot was originally French, from the
Nahuatl word *ocelotl*, a jaguar. Inhabiting Central and South America.

Tiger Cat   *F. tigrina*
tigris (L) a tiger   *-inus* (L) suffix meaning like, pertaining to. Inhabiting Central and South America.

Jaguarundi   *F. yagouaroundi*
Jaguarundi was originally a Tupi name, but this animal is not a jaguar
in spite of the name, and looks nothing like one. It is a rather unusual
member of the cat family and not even like a cat in appearance
Inhabiting Central and South America.

Puma   *F. concolor*
concolor (L) of the same colour (as opposed to *discolor* (L) of different
colours). Adult pumas are plain greyish-brown with no conspicuous
markings. From Canada, USA and South America. It is also known
as the Cougar or Mountain Lion.

Clouded Leopard   *Neofelis nebulosa*
neos (Gr) new; can mean unexpected or strange; zoologists have not
found it possible to classify this leopard as either *Felis* or *Panthera*, and
so have given it a new genus *Neofelis*   *nebula* (L) cloud, hence *nebulosa*
(L) cloudy; the coat is mottled and striped with black on a brownish
grey back-ground. Inhabiting south-east Asia, Borneo, Sumatra and
Java.

Lion   *Panthera leo*
panthera (L) a panther or a leopard (see Leopard)   *panthēr* (Gr)
panther   *leo* (L) a lion. Widespread in Africa except in the north and
a few still remain in India under protection.

Tiger   *P. tigris*
tigris (L) a tiger. Now found only in Asia, Sumatra and Java.

Leopard   *P. pardus*
pardus (L) a panther or a leopard. It is easy to become confused by
these various names, so let us be quite clear that a leopard *is* the same
animal as a panther. The two names may have been used to distinguish between the various sizes and colours that occur in this animal
and 'panther' is usually used for the black variety. Clearly the English

word leopard comes from the two Latin words *leo* and *pardus*. Widespread in Africa, apart from desert areas, and in Asia from the southwest to China and Korea.

Jaguar   *P. onca*

onca   the Greeks were familiar with a moderate sized feline and called it *lynx* (see page 177). The Romans borrowed the Greek word and it became the Tuscan *lonza*. Later the word passed into French as *lonce*, and the initial *l* being mistaken for the article it was elided and the word became *once*, whence it passed into Spanish as *onca* and English as *ounce*. Finally, *onca* was Latinised to *uncia*, although this was Latin for a measure of weight and not an animal! The jaguar inhabits the forests of North, Central and South America.

Snow Leopard or Ounce   *P. uncia*

uncia, see above. Inhabiting the mountains of Central Asia.

Cheetah   *Acinonyx jubatus jubatus*

akaina (Gr) a thorn, a goad   *onux* (Gr) a claw; a reference to the non-retractile pointed claws; this is the recognised explanation but I suggest a possible interpretation would be *a-* (Gr) not   *kineō* (Gr) I move + claw; 'non-moving claws'   *jubatus* (= *iubatus*) (L) maned; young cheetahs have a crest or mane on the shoulders and back. Cheetah is from *cital* (Hind) from *chitraka* (Sanskrit), having a speckled body. Inhabiting Africa south of the Sahara and south-west Asia, it is rare and in danger of extinction.

Indian Cheetah   *A. j. venaticus*

venaticus (L) belonging to hunting; for hundreds of years the human race has tamed and trained cheetahs to assist them in hunting. This subspecies is very rare.

18   Seals, Sea-Lions and the Walrus
PINNIPEDIA

This group, like the whales, consists of mammals that have become adapted to a life in the water. However, unlike the whales they are able to come ashore and move about on land, albeit in rather a slow and ungainly fashion. They are all carnivorous, and in addition to fish they eat a variety of other foods, such as sea birds, shellfish and other small marine life.

Subclass   EUTHERIA
Order   PINNIPEDIA

pinna (L) a feather or wing; can mean a fin   *pes* (L), genitive *pedis*, a foot; 'fin-footed'; they walk on their fins when they come ashore.

Family   OTARIIDAE   13 species

us (Gr), genitive *ōtos*, an ear, thus *otarion*, a little ear; unlike the common seals, of the family Phocidae, they have external ear-flaps.

Cape Fur Seal   *Arctocephalus pusillus*
arktos (Gr) a bear   *kephalē* (Gr) the head; it has a bear-like appearance *pusillus* (L) very small. Fur Seals are so-named because the hide makes

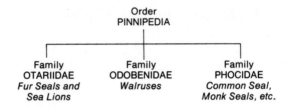

Order
PINNIPEDIA

| Family OTARIIDAE Fur Seals and Sea Lions | Family ODOBENIDAE Walruses | Family PHOCIDAE Common Seal, Monk Seals, etc. |

excellent fur coats; *pusillus* because the description was based on a picture of a young pup. Inhabiting the sea round The Cape of Good Hope, South Africa.

Forster's Fur Seal *A. forsteri*
Named after J. G. A. Forster, FRS (1754-1794), an artist who accompanied Captain James Cook on his second voyage of exploration in *H.M.S. Resolution* during the years 1772 to 1775, and he made a drawing of this seal. It lives in the sea round the coast of New Zealand.

Australian Fur Seal *A. doriferus*
dora (Gr) a skin, a hide; the word only applies when the skin is removed for the fur *fero* (L) I bear. From the southern coasts of Australia and Tasmania, it has particularly fine fur.

Kerguelen Fur Seal *A. tropicalis*
tropikos (Gr) tropical *-alis* (L) suffix meaning relating to; the name was given mistakenly as it does not inhabit tropical seas. Kerguelen Island lies well to the south of the Indian Ocean; it also inhabits other sub-antarctic islands.

South American Fur Seal *A. australis*
australis does not mean Australia; *auster* (L) the south *-alis* (L) suffix meaning pertaining to *australis* (L) southern. It inhabits most of the coastal sea round South America.

Guadalupe Fur Seal *A. philippii*
Dr R. A. Philippi (1808-1904) was at one time Director of the Museum in Santiago in Chile. He collected the skull of this seal in 1864. It inhabits one small protected breeding colony on Guadalupe Island, off the coast of California.

Northern Fur Seal *Callorhinus ursinus*
kallos (Gr) a beauty, a beautiful object *rhinos* (Gr) the skin or hide

of an animal; 'beautiful hide'. T. S. Palmer and others have translated this as *rhinos* (Gr) the nose, and admit to not knowing the
explanation; the Greek for nose is *rhis*, genitive *rhinos*. The Greek
word *rhinos* also means the skin or hide of beasts and is frequently used
in this sense in Homer; I offer this as a more likely interpretation
ursus (L) a bear   *-inus* (L) suffix meaning like; 'bear-like'; it is sometimes known as the Sea-Bear. Inhabiting Pribilof and other islands
in the Bering Sea area of the northern Pacific Ocean.

Californian Sea Lion   *Zalophus californianus*

za- (Gr) an intensive prefix   *lophos* (Gr) a crest; 'a high crest'; there
is a high sagittal crest on the male's skull. This is the sea lion which
you will have seen performing in a circus. It is not confined to the
Californian area; it also inhabits the western side of the Pacific north
of Japan.

Steller's Sea Lion   *Eumetopias jubatus*

eu- (Gr) well, typical   *metopion* (Gr) the forehead; it has a broad forehead   *jubatus* (=*iubatus*) (L) having a mane; the male has a shaggy
mane. G. W. Steller (1709–1746) was a German zoologist who was
exploring in the northern Pacific Ocean in 1740. This sea lion inhabits coastal areas of both sides of the northern Pacific.

Southern Sea Lion   *Otaria byronia*

Otaria, see under Family   Commodore John Byron was serving with
HMS Tamar on a voyage of discovery in the South Seas from 1764 to
1766; he brought home a skull of this sea lion. It inhabits coasts of
South America and the Falkland Islands.

Australian Sea Lion   *Neophoca cinerea*

neos (Gr) new   *phōkē* (Gr) a seal; indicating a newly named sea lion
inis (L) ashes, hence *cinerea* (L) ash-coloured. Inhabiting coasts of
southern Australia.

Hooker's Sea Lion   *Phocarctos hookeri*

phōkē (Gr) a seal   *arktos* (Gr) a bear; from the skull which is like that
of a bear. Sir J. D. Hooker (1817–1911) was a famous English naturalist and travelled widely as the botanist on expeditions in Australia,
Tasmania and New Zealand. This sea lion inhabits the southern
coast of New Zealand.

Family ODOBENIDAE   1 species and 1 subspecies

dous (Gr) a tooth   *bainō* (Gr) I step, I walk; 'one that walks with its teeth'; the walrus has been seen to drag itself along the ice using its tusks which are actually over-developed canine teeth.

Atlantic Walrus or Sea Cow *Odobenus rosmarus rosmarus*
Rossmaal or *rossmar* are Scandinavian names for the walrus; it lives mostly in Arctic areas, from Canada to northern Russia.

Pacific Walrus or Sea Cow *O. r. divergens*
di- (L) prefix, two; can mean apart *vergo* (L) I turn *divergens* (New L) turning apart; referring to the tusks. The Old English name was *horschwael*, meaning 'horse-whale'; it is also sometimes called the morse, from *mors* (Lappish) a walrus. This walrus lives in the area round Alaska, the eastern coasts of Siberia and the Bering Sea.

Family PHOCIDAE 18 species
phoca (L) a seal, a sea-calf.

Common Seal *Phoca vitulina*
vitula (L) a calf *-inus* (L) suffix meaning like. Living on the shores of the northern oceans, including the British Isles.

Ringed Seal *Pusa hispida*
Pusa is the Greenlandic name for a seal; it is probably a misspelling as the Greenlandic name for the harp seal is *puirse*; *hispidus* (L) hairy bristly; a reference to the whiskers: the body is marked with oval white rings. Inhabiting Arctic coasts.

Baikal Seal *P. sibirica*
-icus (L) suffix meaning belonging to; living in Lake Baikal, Siberia a huge fresh-water lake some 640 km (400 miles) long; probably the only seal living entirely in fresh water.

Caspian Seal *P. caspica*
-icus (L) suffix meaning belonging to; of the Caspian Sea.

Harp Seal *Pagophilus groenlandicus*
pagos (Gr) anything hardened, ice, frost *philos* (Gr) loving, fond of they live among ice and snow *-icus* (L) suffix meaning belonging to of Greenland; they are not confined to this area, being widespread in the Arctic. They have unusual and distinct markings on the back which have given rise to the names 'Harp' and 'Saddlebacked'. It is also sometimes known as the Greenland Seal.

Grey Seal   *Halichoerus grypus*
halios (Gr) from, belonging to the sea   *khoiros* (Gr) a pig; 'a sea-pig'
grupos (Gr) hook-nosed; in profile the nose is distinctly rounded.
Inhabiting coastal areas of northern oceans including the British
Isles; it is said that in autumn there are more living round our shores
than all other areas put together. The other 'British' seal is the
Common Seal *Phoca vitulina*.

Crabeater Seal   *Lobodon carcinophagus*
lobos (Gr) a lobe   *odōn* (Ionic Gr) a tooth; the molars are compressed
and have lobes both in front and behind   *karkinos* (Gr) a crab
phagein (Gr) to eat; they probably eat more krill than crabs and this
may be the reason for the shape of the teeth. They live in the Antarctic
area.

Ross's Seal   *Ommatophoca rossi*
omma (Gr), genitive *ommatos*, the eye   *phoca* (L) a seal; it has exceptionally large eyes. Rear Admiral Sir James Ross (1800–1862), a
famous British Arctic and Antarctic explorer, also discovered Ross's
Gull *Rhodostethia rosea*. The gull lives in the Arctic whereas the seal
lives in the Antarctic.

Leopard Seal or Sea-Leopard   *Hydrurga leptonyx*
hudōr (Gr) water; in composition the prefix *hudro-* is used   *ergō* (Gr)
I work; a misspelling, or possibly derived from *urgeo* (L) I drive, I
urge; a reference to its aquatic life   *leptos* (Gr) slender, thin   *onux*
(Gr) a claw, nail. The skin is usually dull brown with light spots;
some have been caught with colouring remarkably like a leopard;
they are also ferocious predators. Inhabiting southern oceans.

Weddell's Seal   *Leptonychotes weddelli*
leptos (Gr) slender, thin   *onux* (Gr), genitive *onukhos* a claw, nail
-otēs (Gr) suffix denoting possession; referring to the rudimentary
claws of the hind feet. This seal was discovered by James Weddell,
the Scottish sealer, and named in his honour. He brought home a
drawing and a skeleton in 1824. The Weddell Sea, discovered in
1823, also bears his name; it is in the Antarctic, where this seal lives.

West Indian Monk Seal   *Monachus tropicalis*
monakhos (Gr) a monk, solitary; they are not always solitary and may
live in colonies in certain areas; they have rings of fat round the neck
which might suggest a monk's hood or cowl   *-alis* (L) suffix meaning

pertaining to; of the tropics. They live in the Caribbean Sea area and the Gulf of Mexico, but are now very scarce and may even be extinct.

Hawaiian Monk Seal *M. schauinslandi*
Professor H. H. Schauinsland (1857–1937), a German zoologist and at one time Director of the Bremen Museum, discovered this seal on Laysan Island, Hawaii. There are probably no more than 1,000 to 1,500 still in existence.

Hooded Seal or Crested Seal *Cystophora cristata*
kustis (Gr) a bladder; also a bag or pouch *phoros* (Gr) carrying *cristata* (L) crested; the male has a peculiar pouch forming a crest on the nose which can be inflated like a bladder. Inhabiting large areas round Greenland in the Arctic.

Southern Elephant Seal *Mirounga leonina*
mirounga (New L) derived from *miouroung*, an Australian native name for the seal *leo* (L), genitive *leonis*, a lion *-inus* (L) suffix meaning like; leonine. A very big seal which can weigh up to four tonnes, with a trunk-like snout; it can make a roaring noise said to be like that of a lion. Widespread in the southern oceans including the coast of Australia.

Northern Elephant Seal *M. angustirostris*
This seal lives on the western coast of North America, and nowhere near Australia, but it takes the Latin name *Mirounga* from the Southern Elephant Seal (above) *angustus* (L) narrow *rostrum* (L) the snout; the male has a snout rather like an elephant's trunk and up to 600 cm (2 ft) long. However, it is narrower than the snout of the southern species *M. leonina*.

All by itself, this peculiar termite-eating mammal constitutes the order Tubulidentata, and the family Orycteropodidae. Why, the reader may ask, does it stand so isolated from other mammals? The reason is that the anatomical structure is quite unlike any other mammal, and gives no indication of the group in which it should be placed. The teeth are simple cylinders of dentine traversed from base to crown by hundreds of minute passages, or tubules. It has no immediate ancestors but probably stems from very early ungulates, as did the hyraxes.

Subclass　EUTHERIA
Order　TUBULIDENTATA
tubus (L) a pipe, a tube　*-ulus* (L) diminutive suffix　*dens* (L), genitive *dentis*, a tooth　*-atus* (L) suffix meaning provided with (see introductory note above).

Family　ORYCTEROPODIDAE　1 species
orukter (Gr) a tool for digging　*pous* (Gr), genitive *podos*, a foot.

Aardvark　*Orycteropus afer*
'One that has feet for digging'; it has powerful claws and is a remarkably fast digger　*Afer* (L) African. Aardvark was originally South

Order
TUBULIDENTATA
|
Family
ORYCTEROPODIDAE
The Aardvark

African Dutch, now known as Afrikaans, for earth-pig. Though rarely seen, being nocturnal, it is widespread in Africa south of the Sahara.

There are now only two species of elephants, the African and the Indian. There is only one family; their exceptional tusks, trunk and many other characteristics set them apart from all other mammals. The trunk has evolved from a shorter proboscis such as that of the tapirs, and has become a specialised organ capable of picking up food, and drawing up water, and various other uses. It is thought that they may be distantly related to the manatees, and represent the only survivors of several hundred extinct species known from fossils.

Subclass EUTHERIA

Order PROBOSCIDEA

pro- (Gr) prefix meaning before, in front of; also used to express motive *boskō* (Gr) I feed, hence *proboskis* (Gr) a trunk or proboscis used for feeding as with elephants and certain insects *idea* (Gr) appearance; can mean a kind, a sort.

Family ELEPHANTIDAE 2 species
elephantus (L) the elephant.

Order
PROBOSCIDEA
|
Family
ELEPHANTIDAE
Elephants

African Elephant   *Loxodonta africana*
loxos (Gr) slanting   *odous* (Gr), genitive *odontos*, a tooth; the grinding
surfaces of the teeth appear to be 'lozenge-shaped'.

Indian Elephant   *Elephas maximus*
elephas (Gr) the elephant   *maximus* (L) largest; this is misleading as
the African species is usually the bigger. It is not confined to India
and inhabits Sri Lanka, forested areas of south-east Asia and Sumatra

21 **Hyraxes** HYRACOIDEA

The hyraxes are small rabbit-like animals. They have always been a problem for zoologists as there seems to be no obvious group in which to place them for purposes of classification. For many years they were thought to be related to the elephants, but recently some zoologists have come to the conclusion that this may not be the case. Because of the unusual anatomical structure, particularly with regard to the teeth and the feet, they must be classed as a separate order. The upper cutting teeth, the incisors, are to some extent rodent-like, but the upper cheek teeth are like those of a rhinoceros, and the lower cheek teeth like those of a hippopotamus. The toes have small hoofs, and on the sole there is what appears to be a sort of suction pad which may account for their agility when climbing rocks and trees. They are the last survivors of about twelve extinct genera.

Subclass **EUTHERIA**
Order **HYRACOIDEA**

hurax (Gr), genitive *hurakos*, a mouse, a shrew-mouse -*oidea* (New L), from *eidos* (Gr) form, sort, a particular kind; it is much bigger than a mouse.

Family **PROCAVIIDAE** about 10 species (authorities differ)

pro- (Gr) before, hence *prōtos* (Gr) first *cavia* (New L) from cabiai, a Brazilian word used in South America for a rodent, probably a guinea-pig'; *Procavia* could mean 'first guinea-pigs', suggesting that other similar animals are descended from these.

Order
HYRACOIDEA
|
Family
PROCAVIIDAE
Hyraxes

Tree Hyrax   *Dendrohyrax arboreus*
dendron (Gr) a tree   *hurax* (Gr) a mouse, a shrew-mouse; it is much
bigger than a mouse, being more like a large guinea-pig, or a rabbit
arbor (L) a tree, hence *arboreus* (L) relating to trees. Inhabiting most
of Africa south of the Sahara.

Beecroft's Hyrax   *D. dorsalis*
dorsum (L) the back   *-alis* (L) suffix meaning relating to; there are
scent glands on the back covered by a patch of white hair; the hair
stands erect when the animal is roused. John Beecroft was an English
naturalist who was made Governor by the Spaniards of their island
Fernando Po, in 1844; five years later he became the British Consul
This hyrax was first found on the island about 1850, probably by
Beecroft. It inhabits the western part of Africa.

Yellow-spotted Hyrax   *Heterohyrax brucei*
heteros (Gr) different + hyrax; an allusion to the skull, which is like
that of *Dendrohyrax* except that the orbit, or eye-socket, is incomplete
behind; James Bruce (1730–1794), a naturalist who was in Ethiopia
(formerly Abyssinia) from 1768 to 1773, made some expeditions t
discover the sources of the Nile. At the time his amazing travel stories
about Abyssinia were not believed. This hyrax inhabits most of Afric
south of the Sahara.

Rock Hyrax or Dassie   *Procavia capensis capensis*
Procavia. see Family   *-ensis* (L) belonging to; it inhabits the Cap
Province area and Namibia. Dassie is the Afrikaans name for th
hyrax.

Syrian Hyrax   *P. c. syriacus*
-acus (L) suffix meaning relating to; this hyrax ranges through Syr
and Sinai and south to Saudi Arabia. It is the 'Coney' mentioned i
the Bible, but it is not a rabbit. A substance known as hyraceum, sa
to be excreted by the hyrax and presumably its droppings, was use
as a medicine, a 'folk remedy', and was supposed to be good f
epilepsy and convulsions!

The Manatee is probably the origin of the mermaid legend, and this may have come about because they have a habit of standing up in the water with their head and shoulders above the surface. When doing this, the females often hold a baby manatee in one flipper to enable it to suckle at the breast.

The Manatee and the Dugong are a puzzle for the zoologists, as among the mammals, for anatomical reasons, it has been necessary to place them in a group by themselves. They are almost hairless and live entirely in the water, but unlike the whales they are completely herbivorous. The hind limbs have evolved into a fleshy horizontal paddle. The general opinion is that they are very distantly related to the elephants. They are the last survivors of about twenty fossil genera.

Subclass **EUTHERIA**

Order **SIRENIA**

irēn (Gr) a siren; according to the legend the Sirens were nymphs who allured sailors by their sweet songs and then slew them.

Family **DUGONGIDAE** 1 species

Duyong is a Malayan name for this marine animal.

Dugong or Halicore *Dugong dugon*

The name halicore is from *hals* (Gr) the sea, and *korē* (Gr) a girl; like the manatee it is supposed to have given rise to the mermaid legend. It is found along the coasts round the Indian Ocean, and ranging south to Indonesia and northern Australia; it is becoming very rare.

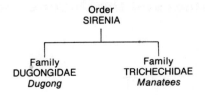

Family   TRICHECHIDAE   3 species
thrix (Gr), genitive *trikhos*, hair   *ekhō* (Gr) I have; they have hair on the face and bristly moustaches.

North American or Caribbean Manatee   *Trichechus manatus*
manati (Sp) the manatee, derived from a native name in Haiti. Inhabiting the coast of Haiti and other islands in the Caribbean Sea.

South American Manatee   *T. inunguis*
in- (L) prefix meaning not, without   *unguis* (L) a claw, a hoof; the flippers are paddle-like and without claws. Inhabiting the Atlantic coasts of both North and South America.

West African Manatee   *T. senegalensis*
-ensis (L) suffix meaning belonging to. It is not confined to the Senegal area of West Africa and may also travel far inland up the rivers. These strange animals are sometimes known as Sea Cows.

Horses, Tapirs and Rhinoceroses
PERISSODACTYLA

This group, together with the larger group comprising the antelopes, pigs, cattle and their kin, are known as ungulates, though the term is not usually employed in classification; it means hoofed, from the Latin *unguis*, a claw or hoof.

The members of the group we are now going to deal with have an uneven number of toes or hoofs and have been given the name Perissodactyla, meaning 'odd-toed'. They have either one or three toes, though the tapirs are an exception, having three toes on the hind feet and four on the front feet. The rhinoceroses have three toes on each foot.

In all cases, the central axis of the foot passes through the third digit, which may take most of the weight, and is large and symmetrical. In the case of the horses, asses and zebras, it is the only digit remaining, taking all the weight; the others have disappeared during evolution, a process occupying some 18,000,000 years, though rudimentary remains can still be seen. Thus, they walk and run on the middle finger!

For purposes of classification, the order is divided into two suborders, namely Hippomorpha, the 'horse-kind', and Ceratomorpha, the 'horned-kind'. The antelope group, having an even number of toes or hoofs, have been given the name Artiodactyla, meaning 'even-toed'; this will be dealt with in Chapter 24.

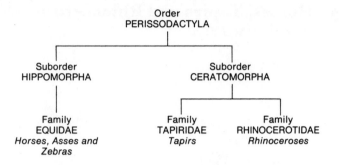

Order PERISSODACTYLA
Suborder HIPPOMORPHA

perissos (Gr) strange, unusual; of numbers odd, uneven   *daktulos* (Gr)
a finger; can mean a toe.

hippos (Gr) a horse   *morphē* (Gr) apparent form, a kind, sort; the
'horse-kind'.

Family EQUIDAE   about 6 species
equus (L) a horse

Wild Horse   *Equus przewalskii*
General N. M. Prjevalski (1839–1888) was a famous Russian explorer
and naturalist who made several expeditions to Central Asia; he
collected thousands of birds as well as mammals. Sometimes known
as Prjevalski's Horse it is very rare, but still inhabits the Altai Mountains in Western Mongolia. It is the only true wild horse known to be
still in existence.

Horse   *E. caballus*
caballus (L) a pack horse, a nag; this is the domestic horse.

Central Asian Ass   *E. hemionus hemionus*
hēmi- (Gr) prefix meaning half   *onos* (Gr) an ass   *hēmionos* (Gr)
mule; the name is misleading, as a mule is the offspring of a male ass
mating with a mare, and it would be sterile and could not breed. In
this case *hēmi* is meant to indicate 'like an ass' rather than 'half-ass'

It has a variety of different local names depending on where it is found; in Mongolia it is known as the Dziggetai.

Onager or Persian Wild Ass   *E. h. onager*
onos (Gr) an ass   *agrios* (Gr) wild, savage   *onagrus* or *onager* (L) a wild ass. In India it is known as the *Ghor-kar*.

Kiang   *E. kiang*
Kiang or *Kyang* is a Tibetan native name; this wild ass lives in Tibet, western China. Some give it as a subspecies of *E. hemionus*.

African Wild Ass   *E. africanus*
Generally accepted as the ancestor of the domestic donkey (below). It inhabits northern parts of Africa but is becoming very rare.

Domestic Donkey   *E. asinus*
asinus (L) an ass; the name was given by Linnaeus in 1758.

Cape Mountain Zebra   *E. zebra zebra*
zebra is obscure, but could be from a Congolese or possibly an Abyssinian word *zibra*, meaning striped. Inhabiting the southern mountains of South Africa, it is very rare, and found only in certain protected areas.

Hartmann's Zebra   *E. z. hartmannae*
Premier Lieutenant Dr Hartmann discovered this zebra on the Kaoko Veld plateau, Namibia, and sent two skins to the Berlin Museum as a gift. It was named in 1898 by Dr Paul Matschie, of the Berlin Museum, in honour of Mrs Hartmann. Inhabiting the mountains of South West Africa.

Common Zebra   *E. burchelli*
Dr W. J. Burchell (1782–1863) was an English zoologist who made an expedition to Africa in 1811; this zebra is widespread in Africa south of the Sahara. The nominate subspecies, that is the animal originally described in 1824 as *E. burchelli*, from the northern Cape Province and the Orange Free State, and now known as *E. b. burchelli*, has been extinct for many years.

Chapman's Zebra   *E. b. antiquoram*
antiquorum (L) the ancients, of old times. It was named after James Chapman, an English naturalist who travelled in Africa in the nineteenth century. Sometimes known as the Damaraland Zebra, it inhabits the southern part of Africa, ranging from Angola to the Transvaal.

Selous's Zebra   *E. b. selousi*
F. C. Selous (1857–1917) was a big game hunter in Africa from 1871 to 1917. This zebra lives in the southern part of Rhodesia, Malawi, and the area between.

Grant's Zebra   *E. b. boehmi*
Named after Colonel J. A. Grant (1827–1892) who explored Central Africa during the years 1860 to 1862, and Dr R. Böhm (1854–1884), a German zoologist who was in Africa during the years 1880 to 1884. Probably the commonest zebra, it ranges from Sudan and Ethiopia south to Uganda, Kenya, Tanzania and Zambia.

Grevy's Zebra   *E. grevyi*
Named after Francois P. J. Grevy (1807–1891), who was President of the French Republic from 1879 to 1887. This zebra is easily distinguished by its large rounded ears and unusually narrow stripes; it is interesting to note that since November 1912 it has been given five different generic or subgeneric names, and four of these during the course of one year! It inhabits Ethiopia, Somaliland, and northern Kenya.

Suborder   CERATOMORPHA

keras (Gr), genitive *keratos*, a horn   *morphe* (Gr) form, shape; can mean kind, sort; the 'horned-kind'; however it is only the rhinoceroses that possess the nose-horn.

Family   TAPIRIDAE   4 species
Tapir is from the Tupi word *tapira*; the Tupis are a tribe of South American aborigines living in the Amazon area.

Malayan Tapir   *Tapirus indicus*
Indicus (L) of India; the name is misleading as this tapir does not live in India. When Professor Desmarest named it in 1819, he probably meant to indicate the East Indies, as it lives in the Malay Peninsula and Sumatra.

Brazilian Tapir   *T. terrestris*
terrestris (L) of the earth, land-dwelling. Inhabiting a large area in Brazil and also Venezuela and other areas in the north.

Mountain or Woolly Tapir   *T. pinchaque*
La Pinchaque was a large fabulous animal believed to live in the same range of mountains as this tapir, which inhabits the Cordillera

Occidental, a range of mountains in the north-west of Colombia in South America.

Baird's Tapir *T. bairdi*
W. M. Baird was an American naturalist who made an expedition to Mexico in 1843. In fact, previous accounts of the animal had been recorded by W. T. White, another American naturalist. It lives in Mexico, Central America, and ranges south to Ecuador.

Family RHINOCEROTIDAE 5 species
rhis (Gr), genitive *rhinos*, the nose   *keras* (Gr), genitive *keratos*, a horn.

Indian Rhinoceros *Rhinoceros unicornis*
unus (L) one   *cornu* (L), genitive *cornus*, the horn of an animal; there is only one horn. It inhabits limited areas in Bengal, Assam and Nepal, and is now very rare.

Javan Rhinoceros *R. sondaicus*
-icus (L) suffix meaning belonging to; the Malay islands are known as the Sunda Islands and the Sunda Strait lies between Java and Sumatra, hence *sondaicus*. This rhino is very rare and is confined to a small area in Java.

Sumatran Rhinoceros *Didermocerus sumatrensis*
di- from *dis* (Gr) two, double   *derma* (Gr) skin   *keras* (Gr) a horn; the horn is composed of keratin, a substance derived from the skin, like finger-nails; there are two horns   *-ensis* (L) suffix meaning belonging to; it is not confined to Sumatra and was once widespread in south-eastern Asia, from Assam to Malaya. Like the other rhinos from this part of the world, it is very rare.

White Rhinoceros *Diceros simus*
Diceros, see above   *simum* (L) snub-nosed, flat-nosed; it has a flat, wide nose and lips adapted for grazing. The white rhino is not white, and the black rhino is not black; they are both grey; the name may be a corruption of the Dutch word *weit*, meaning wide, and referring to the nose and lips. The numbers of all species have been seriously reduced because they are hunted for the horn, which is supposed to be an aphrodisiac when ground into a powder. It inhabits two separate areas of Africa, one in the south, and one central, to the north-west of Lake Victoria.

Black Rhinoceros   *D. bicornis*

Diceros, see above   *bi-* from *bis* (L) twice, two   *cornu* (L), genitive *cornus*, the horn of an animal; so here it is first in Greek for the genus, and then in Latin for the species; 'a two-horned two-horn'. This rhino is dark grey, not black, and has a narrow rather pointed nose and lips, adapted for browsing; see notes re White Rhinoceros, above. The Black Rhinoceros ranges over a wide area in the southern half and the eastern side of Africa.

Pigs, Camels, Deer, Giraffes, Antelopes and their kin ARTIODACTYLA

This group, together with the smaller group comprising the horses, zebras, rhinoceroses and their kin, are known as ungulates, though the term is no longer in regular use in classification. It means 'hoofed', from the Latin *unguis* a claw or hoof. They have an even number of toes or hoofs and so have been given the name Artiodactyla, meaning 'even-toed'; the main axis of the foot passes between the third and fourth digits, which are capped with hoofs. In most cases, these two digits, forming the 'cloven hoof', take all the weight, the others having disappeared during the course of evolution. In some cases, however, such as the hippopotamuses and the chevrotains, four digits are still in use, or at least visible.

The horses and rhinoceroses, already dealt with in Chapter 23, have an uneven number of toes or hoofs and so have been given the

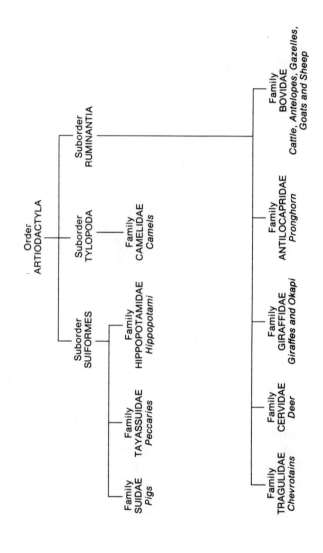

name Perissodactyla, meaning 'odd-toed'. All the animals in these two groups are related though very different in appearance.

Among the animals that follow will be found most of those that man has domesticated. Throughout the world, many millions of these are in use today for supplies of food, milk, wool, leather and other purposes. It is interesting to note that they are nearly all vegetarians and yet they are our main supply of meat.

For purposes of classification the order is divided into three sub-orders, namely Suiformes, meaning 'pig-like'; Tylopoda, meaning 'pad-footed'; and Ruminantia, meaning 'cud-chewers'.

Subclass EUTHERIA

Order ARTIODACTYLA

artios (Gr) complete, perfect of its kind; of numbers, even   *daktulos* (Gr) a finger; can mean a toe.

Suborder SUIFORMES

sus (L), genitive *suis*, a pig   *forma* (L) shape, sort, kind; the 'pig-kind'.

Family SUIDAE   8 or 9 species

African Bush Pig   *Potamochoerus porcus*
potamos (Gr) a river   *khoiros* (Gr) a pig; sometimes known as the River Hog, they are fond of wallowing in water in hot weather *porcus* (L) a pig or hog. Like most other pigs, it is not entirely vegetarian; the range is widespread in Africa south of the Sahara, and it is also found in Madagascar, where it is the only 'Ungulate'.

Giant Bornean Pig   *Sus barbatus*
sus (L) a pig   *barbatus* (L) bearded; it has a moustache, rather than a beard. Inhabiting Borneo, the Philippines and Malaya.

Eastern Wild Boar   *S. verrucosus*
verruca (L) a wart   *-osus* (L) suffix meaning 'full of'; it has three warts on each side of the head and muzzle. It inhabits Java.

European Wild Boar   *S. scrofa*
scrofa (L) a breeding sow; this is the boar from which domestic pigs have been derived. It has a wide distribution, including Europe, North Africa, Asia, Sumatra, Java, Formosa and Japan.

Pygmy Hog   *S. salvanius*
salvanius, belonging to Saul Forest in Nepal. The name was given by

Hodgson in 1847. Inhabiting Nepal, Sikkim and Bhutan. It was thought to be extinct but some specimens were found in 1971.

Wart Hog   *Phacochoerus aethiopicus*
phakos (Gr) a mole or wart on the body   *khoiros* (Gr) a pig or hog; it has warts of various sizes on both sides of the face   *aethiopicus* is often used to indicate Africa as a whole. This hog is widespread in Africa south of the Sahara.

Meinertzhagen's Giant Forest Hog   *Hylochoerus meinertzhageni*
hulē (Gr) a wood, forest   *khoiros* (Gr) a pig or hog, Colonel R. Meinertzhagen (1878–1967), the naturalist who discovered this hog, was in East Africa from 1902 to 1905. Like most other hogs it is not entirely vegetarian; it inhabits forested areas of the central part of Africa.

Family   TAYASSUIDAE   2 species
tajacu or *tayassu* is a Brazilian native name for the peccary.

Collared Peccary   *Tayassu tajacu*
Pecary was originally a Brazilian native (Tupi) name, and said to mean 'many paths through the woods', a reference to the animals habit. It has a narrow collar of light hair on the shoulders and is an agressive, fierce, hog-like animal with scimitar-like tusks. It will attack most living things and is not entirely vegetarian. Inhabiting Arizona, New Mexico, Texas, Mexico, Central America and the northern part of South America.

White-lipped Peccary   *T. albirostris*
albus (L) white   *rostrum* (L) the snout; it has white lips and a white moustache. Inhabiting an area similar to the Collared Peccary (above).

Family   HIPPOPOTAMIDAE   2 species
hippos (Gr) a horse   *potamos* (Gr) a river.

Hippopotamus   *Hippopotamus amphibius*
amphi (Gr) around, on both sides   *bios* (Gr) living, manner or means of living; 'living on both sides', i.e. land and water. In a strict sense the hippo is not an amphibian as that sort of animal starts life equipped with gills, like a fish. The hippopotamus is widespread in the rivers and lakes of the southern half of Africa, but not now in the most southern part.

Pygmy Hippopotamus   *Choeropsis liberiensis*
khoiros (Gr) a pig or hog   *opsis* (Gr) aspect, appearance   *-ensis* (L) suffix meaning belonging to; not confined to Liberia, it also inhabits Guinea and Sierre Leone, but is now very rare.

Suborder   TYLOPODA
.tulē (Gr) a swelling or lump, a pad for carrying burdens   *pous* (Gr), genitive *podos*, a foot; 'pad-footed' referring to the tough padded soles of the feet.

Family   CAMELIDAE   6 species
camelus (L) a camel.

Arabian or One-humped Camel   *Camelus dromedarius*
A dromedary is really a domesticated camel bred for travel, for running. The name comes from *dromeus* (Gr) a runner, and *-arius* (L) a suffix meaning pertaining to, hence *dromedarius* (New L) a running camel. It is unknown except in a domesticated state.

Bactrian Camel   *C. bactrianus*
Bactria is a province of the ancient Persian Empire   *-anus* (L) suffix meaning belonging to. This is the two-humped camel, and also domesticated, being almost unknown except in that state. However, there may be some still living in the desert regions of China and Central Asia.

Llama   *Lama peruana* or *L. glama*
Llama is a Peruvian name for this animal, which is closely related to the camel, and probably originates from the Quéchua, an ancient people of Peru   *glama*, a name given by Linnaeus, is a corruption of *lama*. There are no llamas in the wild now, but large herds are domesticated and kept by the South American Indians. Llamas are used as pack animals and kept also for the hide, the fur for rugs and clothing, and the flesh for food.

Guanaco   *L. guanicoe*
Guanaco is a Spanish name from the Quéchua people. This is a slightly different animal from the llama and still survives in the wild, ranging from Peru south to Patagonia in Argentina.

Alpaca   *L. pacos*
Paca and *pacos* are Peruvian names for the llama: *al* (Arabic, adopted in Spanish) means 'the', so alpaca means 'the paca'; the word is used in the English language to mean a cloth made from its wool. It no

longer survives in the wild but probably more than a million domesticated animals are kept in South America for their wool.

Vicuña   *Vicugna vicugna*
Vicuña is a Peruvian name for this animal; it still survives in the wild and lives high up in the Andes in Peru, Bolivia, Chile and Argentina. Like all the animals in this family, it is camel-like in many ways.

Suborder   RUMINANTIA
rumino (L) I chew the cud.

Family   TRAGULIDAE   4 species
tragos (Gr) a goat   *-ulus* (L) diminutive suffix.

Water Chevrotain   *Hyemoschus aquaticus*
hus (Gr), genitive *huos*, a hog   *moskhos* (Gr) musk; 'hog musk-deer', referring to the type of skull and its pig-like habits   *aquaticus* (L) living in water or near water. It inhabits Liberia and part of Nigeria.

Indian Chevrotain   *Tragulus meminna*
tragos (Gr) a goat   *-ulus* (L) diminutive suffix; it is not a goat, but somewhat resembling a very small deer and structurally more akin to pigs and camels   *memina* (Ceylonese) a small deer. Chevrotain is from the French *chèvre*, a she-goat. It lives in the southern part of Asia.

Larger Malay Chevrotain   *T. napu*
Napu is a native name for the chevrotain in Sumatra. Inhabiting southern Asia, Sumatra and Borneo.

Lesser Malay Chevrotain   *T. javanicus*
-icus (L) suffix meaning belonging to; the range is similar to other chevrotains but it is also found in Java. Sometimes known as the Mouse Deer it is only about 30 cm (1 ft) high.

Family   CERVIDAE   between 30 and 40 species
cervus (L) a stag, a deer.

Musk Deer   *Moschus moschiferus*
moskhos (Gr) musk, *moschus* (New L) musk   *fero* (L) I bear, I carry the male has a pouch on the abdomen and the glands in this pouch produce a strongly scented musk. It is widespread in Asia ranging from the north down to southern China.

Chinese Water Deer   *Hydropotes inermis*
hudōr (Gr) water, *hudr-* as a prefix   *potēs* (Gr) a drinker; referring to
the animal's liking for marshy ground   *inermis* (L) unarmed; it does
not have antlers. Inhabiting eastern Asia, and feral in Great Britain.

Indian Muntjac   *Muntiacus muntjak*
Muntiacus (New L) from *muntjak*, the native name in the Sunda
language, in western Java. It has a wide range, from India to southern
China, Burma, Malaya, Sumatra, Borneo and Java; it is sometimes
known as the Javan Muntjac.

Chinese Muntjac   *M. reevesi*
Named after John Reeves, FRS (1774–1856), a British naturalist who
was resident in China during the years 1812 to 1831. He studied the
natural history of the country and sent specimens to England. Reeves's
Long-tailed Pheasant *Syrmaticus reevesi* is also named in his honour.
This muntjac inhabits China, Burma and Thailand, and is feral in
Great Britain.

Black or Hairy-fronted Muntjac   *M. crinifrons*
crinis (L) hair   *frons* (L) forehead, brow. It inhabits south-east Asia.

Tenasserim Muntjac   *M. feae*
Leonardo Fea (1852–1903) was an Italian zoologist who collected in
Burma. Tenasserim is a Division in Burma and there is also a town
and a river with this name. It inhabits south-east Asia.

Roosevelts' Muntjac   *M. rooseveltorum*
The unusual ending of the specific name is the genitive plural, and
refers to the two sons of Theodore Roosevelt, at one time President
of the USA. It inhabits the southern part of Asia.

Tufted Deer   *Elaphodus cephalophus*
elaphos (Gr) a deer   *odous* (Gr) a tooth; referring to the large upper
canines of the male   *kephalē* (Gr) the head   *lophos* (Gr) a crest; the
male has a crest of hair on the forehead and around the base of the
antlers. Inhabiting China.

Fallow Deer   *Dama dama*
dama (*damma*) (L) a general name for animals of the deer kind. Fallow
from Old English *falu*, meaning brownish yellow. Originally from
the Mediterranean region of southern Europe and Asiatic Turkey,
is now widespread in Europe, including the British Isles.

Persian Fallow Deer   *D. mesopotamica*
mesos (Gr) middle, between   *potamos* (Gr) a river; between two rivers,
the Euphrates and the Tigris, hence Mesopotamia, now Iraq. It is
found in a small area in Iran.

Chital or Axis Deer   *Axis axis*
Axis (L) is said to be Pliny's name for this deer, though some records
only show it as 'an unknown wild animal in India'. It inhabits
southern Asia including Sri Lanka. The name chital comes from
cital (Hind) meaning spotted, variegated; it has conspicuous white
spots on a reddish-brown coat. The name cheetah (*Acinonyx jubatus*,
page   ) is from the same source.

Hog Deer   *A. porcinus*
porcus (L) a hog   *-inus* (L) suffix meaning like; referring to its supposed
likeness to a hog. It is found in India, Assam, Burma, Thailand and
Vietnam.

Kuhl's or Bawean Deer   *A. kuhli*
Dr H. Kuhl (1796–1821) was a German naturalist who was in the
East Indies in 1820 and 1821. This deer is confined to Bawean Island
a small island in the Java Sea, between Java and Borneo.

Calamian Deer   *A. calamianensis*
This deer is found only on a group of islands in the Philippines, known
as the Calamian Islands, between Mindoro and Palawan. The species
is in danger of becoming extinct.

Thorold's Deer   *Cervus albirostris*
cervus (L) a stag, a deer   *albus* (L) white   *rostrum* (L) the snout; it
has a white muzzle and lips. Dr W. G. Thorold obtained the second
two specimens in Tibet in 1891. This deer was originally discovered
by Prjevalski in 1879. Inhabiting part of China and Tibet.

Barasingha or Swamp Deer   *C. duvauceli*
A. Duvaucel (1796–1824) was a French naturalist. *Barasingha* is
Hindi (a language of northern India) meaning 'twelve horns', from
bārah, twelve, and *sig*, a horn. This deer usually has a total of twelve
tines on the antlers. Inhabiting India and Assam.

Red Deer   *C. elaphus*
elaphos (Gr) a deer. Inhabiting Europe, including the British Isles
and part of northern Africa and Asia. It is estimated that there are
over 150,000 living in the Highlands of Scotland.

Wapiti   *C. canadensis*
-ensis (L) suffix meaning belonging to; it is not confined to Canada.
Wapiti is an Algonkian name for this large deer; the Algonkin, an
American Indian tribe, still exist today in some reserves in Ontario
and Quebec. The Wapiti inhabits North America on the western side,
and because of its large size it is known there as the Elk; this is
zoologically incorrect and Wapiti is a better name.

Eld's Deer   *C. eldi*
Named after Percy Eld who discovered this 'nondescript' species in
Assam in 1842. Also known as the Thamin, this is a Burmese name
for this deer. It is confined to a small area in southern Burma and in
danger of becoming extinct.

Sika   *C. nippon*
nippon (Jap) Japan   *sika* (Jap) a small deer; the Sika is regarded as
sacred by the Japanese. Now rare, but still found in southern Asia,
Formosa (now Taiwan) and Japan, and feral in Great Britain.

Schomburgk's Deer   *C. schomburgki*
Sir Robert H. Schomburgk (1804–1865) was the British Consul in
Bangkok from 1857 to 1864. This deer is very rare and confined to a
small area in Thailand; it may even be extinct.

Rusa Deer   *C. timorensis*
ensis (L) belonging to; it is not confined to Timor and inhabits
Celebes, Java, Borneo and other islands in that area. *Rūsā* is the Hindi
name for this deer.

Sambar   *C. unicolor*
unicolor (L) of one colour; the coat is a rather dull brown throughout
though there may be some white hair on the neck of the male. *Sambar*
is another Hindi name for this deer; it is widespread in southern Asia.

Père David's Deer   *Elaphurus davidianus*
elaphos (Gr) a deer   *oura* (Gr) the tail; this refers to the unusually long
tail, almost like that of a donkey. Père Armand David (1826–1900)
was a missionary in China and a keen naturalist. He observed this
deer in the Imperial Hunting Park near Peking. Originally from
China it is unknown in the wild though there may be as many as 300
in zoos and wildlife sanctuaries.

Mule or Hollow-toothed Deer   *Odocoileus hemionus*
odous (Gr) a tooth   *koilos* (Gr) hollow; referring to the well-hollowed

teeth   *hēmionos* (Gr) a half-ass, a mule; it has very large ears like a
mule. Inhabiting the western part of North America from Alaska in
the north to Mexico in the south.

White-tailed Deer   *O. virginianus*
-anus (L) suffix meaning belonging to; it is not confined to Virginia
and is widespread in the southern half of North America and the
northern half of South America. In contrast to the black tail of *O.
hemionus* (above), it has a white tail and rump. The tail is raised as a
danger signal when the animal is in flight from predators.

Roe Deer   *Capreolus capreolus*
caper (L) a goat   *-olus* (L) diminutive suffix   *capreolus* (L) the roe
buck. The Roe Deer is widespread in Europe including the British
Isles and ranges to the east through China as far as Korea.

Moose or Elk   *Alces alces*
alces (L) the elk. It is the largest living species of deer and it has been
said the name derives from the Greek *alkē*, meaning strength. Moose
is derived from *musee*, an Algonkian Indian name. Usually known as
a moose in America and an elk in the Old World, it inhabits northern
parts of North America, northern Europe, and northern Asiatic
Russia.

Reindeer   *Rangifer tarandus*
Old French *rangier*, a reindeer, and *ferus* (L) wild, untamed; also said
to come from *ren* (Old Swed) a reindeer, and *ferus*; *tarandrus* (L) a
animal of northern countries, and according to Cuvier the reindeer
It inhabits the tundras and northern woodlands of Europe, Asia
North America and Greenland. Sometimes known as the Caribou

Marsh Deer   *Blastocerus dichotomus*
blastos (Gr) a shoot, or bud   *keras* (Gr) a horn; the horns are said to
resemble a bud   *dikhē* (Gr) in two ways   *tomē* (Gr) cutting, sharp
the small antlers are doubly forked. Inhabiting central Brazil and
south to northern Argentina.

Pampas Deer   *Ozotoceras bezoarticus*
ozōtos (Gr) branched, forked   *keras* (Gr) the horn of an animal; an
allusion to the large, complex, forked antlers   *bezoar* (Sp) a wild goat
-icus (L) belonging to (for meaning of bezoar see Wild Goat, page
224). The Pampas Deer inhabits central Brazil and south to Argentina

Andean Deer or Guemal *Hippocamelus antisensis*
hippos (Gr) a horse   *kamēlos* (Gr) a camel; the scientific name of this
deer has been changed several times in the past. It has been described
as a 'horse-camel'. It really bears no resemblance to either of these
animals, but was considered intermediate between a horse and a
llama. Antisana is the name of a 5,800 m (19,000 ft) peak in the Andes
Mountains where this deer lives   *-ensis* (L) suffix meaning belonging
to   *guemul* or *guemal* is the American Spanish name for this small deer.

Red Brocket *Mazama americana*
Mazame or *maçame* were Mexican names given to some species of deer
in the seventeenth century   *brocart* (Fr) a brocket   *broc* (Old Fr)
tine of a stag's horn. Widespread in Central America and the northern
part of South America.

Little Red Brocket *M. rufina*
rufus (L) red   *-inus* (L) like, pertaining to. Inhabiting South America.

Brown Brocket *M. gouazoubira*
Guazú-birá is a Guarani name in Paraguay for a brocket; a good
example of a 'barbarism' (see page 15). Inhabiting Central and
South America.

Pudu *Pudu pudu*
Pudu is Spanish from the Mapuche, a people of southern Chile. This
very small deer, standing only about 30 cm (12 in) high, is now found
only in the southern Andes in Argentina and on the nearby Chiloe
Island.

Family GIRAFFIDAE 2 species
giraffa (New L) a giraffe, from the Arabic *zarāfah*, meaning 'one who
walks swiftly'.

Okapi *Okapia johnstoni*
Okapi is a native name used by the pygmies in the Semliki Forest area,
where this animal lives. It is related to the giraffe, though very different in appearance and habitat. Sir Harry H. Johnston (1858-1927),
explorer and author, was in the Colonial Administration of British
Central Africa when he discovered the okapi in 1900. A very strange
animal, it was given the name *Equus ? johnstoni*, 'Johnston's Horse'.
However, it soon proved to be nothing like a horse, and was related
to the giraffe, and given the name *Okapia johnstoni*. Sometimes known

as the Forest Giraffe, it is found only in the Semliki Forest area of Zaire.

Giraffe   *Giraffa camelopardalis*
camelus (L) a camel   *pardus* (L) a panther or a leopard   *-alis* (L) relating to, like; this could be interpreted as 'a camel marked like a leopard', but the name does not seem very apt. There is only one species in the genus *Giraffa*, but there is so much variation of colour and marking that a number of subspecies are recognised, most authors giving as many as eight; four well known subspecies are given below. The giraffe is found only in Africa.

Northern Giraffe   *G. c. camelopardalis*
Sometimes known as the Nubian Giraffe, it inhabits Nubia, in Sudan, and parts of Ethiopia.

Reticulated Giraffe   *G. c. reticulata*
reticulum (L) a little net, hence *reticulatus* (L) reticulated, marked like a net; probably the most handsome giraffe.

Masai Giraffe   *G. c. tippelskirchi*
E. L. von Tippelskirch (1774-1840) was a German General and explorer. This giraffe inhabits the Masai Steppe in Tanzania.

Angola Giraffe   *G. c. angolensis*
-ensis (L) suffix meaning belonging to; inhabiting Angola and that area on the western side of South Africa, it is now very rare.

Family   ANTILOCAPRIDAE   1 species
antholops (Late Gr) a horned animal, probably an antelope   *capra* (L) a she-goat. Cuvier suggests that *Antilope* is a corruption of antholops ... 'which seems to refer to the beautiful eyes of the animal'   *antho.* (Gr) a flower, and *ops* (Gr) the eye; '*Ce nom n'est pas ancien, il est corrompu d'antholops ... qui semble se rapporter aux beaux yeux de l'animal.*'

Pronghorn   *Antilocapra americana*
A solitary species and not one of the true antelopes, which are found only in Africa and Asia. The horns are shed every year, as with deer but the bony core remains, as with antelopes. It is the last survivor of an ancient family that lived only in North America. The name pronghorn refers to the short forward-pointing prong on the upper part of the horns. It is now found only in the western part of North America.

Family BOVIDAE

Zoologists do not agree about the number of species; estimates vary from 100 to 154; this is because some have raised certain subspecies to full specific rank, or have decided that certain species are really only subspecies; the latter are not counted when estimating the number of species in a family

bos (L), genitive *bovis*, an ox.

Red-flanked Duiker   *Cephalophus rufilatus*

kephalē (Gr) the head   *lophus* (Gr) a crest; referring to the tuft of hair on the head   *rufus* (L) red   *latus* (L) broad, wide; can mean the side, flank, of men or animals. It inhabits a large area in the central part of western Africa. Duikers are small antelopes, most of them only about 35–45 cm (14–18 in) high   *duiker* is Afrikaans for a diver and supposed to be because they 'dive' into the undergrowth when alarmed.

Blue Duiker   *C. monticola*

mons (L), genitive *montis*, a mountain   *colo* (L) I cultivate; can mean I dwell in a place, I inhabit, hence *monticola* (L) a dweller among mountains. This is the smallest species, being only about the size of a hare; it has a bluish grey coat. Inhabiting a large area of central Africa and ranging south, on the eastern side, down to South Africa.

Zebra Antelope or Banded Duiker   *C. zebra*

zibra (Congolese or possibly Abyssinian) a zebra; can also mean striped; it has a bright orange-coloured coat with black bands or stripes across the back. The rarest species and found only in Liberia and surrounding areas.

Red Duiker   *C. natalensis*

-ensis (L) suffix meaning belonging to; it is not confined to Natal and is fairly widespread on the eastern side of South Africa.

Black Duiker   *C. niger*

niger (L) black; can mean dark, dusky; the coat is dark grey. Inhabiting Liberia, Ivory Coast, Ghana and surrounding areas.

Grey Duiker   *Sylvicapra grimmia*

silva (L) a wood, a forest   *capra* (L) a she-goat; it is not a goat, of course, but a very small antelope like *C. niger* (above) with a blue-grey coat. It is named in honour of Dr Hermann Nicolas Grimm, a

German scientist, who described the duiker as early as 1686, and was given the name by Linnaeus in 1758.

Royal or Pygmy Antelope   *Neotragus pygmaeus*
neos (Gr) new   *tragos* (Gr) a he-goat; when named, in 1827, it was thought to be a new kind of antelope   *pugmē* (Gr) the fist, hence *pugmaios* (Gr) as small as a fist, dwarfish; probably the smallest antelope known, less than 30 cm (12 in) high. On account of its size it became known locally as the King of Hares, hence Royal Antelope. Inhabiting Ghana.

Bates's Dwarf Antelope   *N. batesi*
Named after George Latimer Bates (1883–1940), who settled as a farmer in Cameroun and devoted himself to collecting birds of West Africa. By all accounts he was a retiring and modest man but a most accurate recorder of zoological detail. This very small antelope lives in the Cameroun area of West Africa.

Zanzibar Antelope or Suni   *Nesotragus moschatus*
nesos (Gr) an island   *tragos* (Gr) a he-goat; originally from French Island, or possibly from Chapani Island, close to Zanzibar   *moschatus* (New L) musky; it has musk glands on the feet. It is not confined to any island and can be found in a wide area of territory along the east coast of Africa southward to Mozambique. *Suni* is a native name in south-eastern Africa.

Salt's Dik-dik   *Madoqua saltiana*
Madoqua, from the Amharic (an Ethiopian language) *medaqqwa* a small antelope. Sir Henry Salt (1780–1827) an explorer, was also British Consul General in Alexandria from 1815 to 1827. The name dik-dik is an imitation of the female's cry of alarm. It lives in northern Ethiopia.

Kirk's Dik-dik   *Rhynchotragus kirki* (or *Madoqua*)
rhynkhos (Gr) the snout   *tragos* (Gr) a he-goat; a reference to the trunk-like flexible nose. Sir John Kirk (1832–1922), a Scottish naturalist, was British Consul General in Zanzibar from 1880 to 1887 and was formerly physician and naturalist to Dr Livingstone on his second journey. This dik-dik inhabits East Africa and Namibia.

Klipspringer   *Oreotragus oreotragus*
oros (Gr), genitive *oreos*, a mountain   *tragos* (Gr) a he-goat   *klip* is Dutch for a rock; a 'rock-jumper'; it walks on the tips of its hoofs, and

its agility and sure-footedness on rocks is incredible. Living in mountainous rocky areas throughout Africa.

Beira Antelope   *Dorcatragus megalotis*
dorkas (Gr) a gazelle or antelope   *tragos* (Gr) a he-goat   *megas* (Gr) big   *ous* (Gr), genitive *ōtos*, the ear; the ears are unusually large and wide   *beira* is from *behra*, the Somali name for this antelope. It is a rock-dweller in the mountains of Ethiopia and Somaliland.

Steinbok   *Raphicerus campestris*
raphis (Gr) a needle, a pin   *keras* (Gr) the horn of an animal; the horns are only short spikes   *campestris* (L) level country, a plain. Steinbok is from *steen* (Du) a stone, and *bok* (Du) a buck. Inhabiting South Africa and part of East Asia.

Grysbok   *R. melanotis*
melas (Gr) black   *ous* (Gr) genitive *ōtos*, the ear. Grysbock is from *grys* (Du) grey, and *bok* (Du) a buck; there is white hair scattered in the coat. It is found only in the most southern part of South Africa.

Oribi   *Ourebia ourebi*
The name *oribi* is a Hottentot word, of the Nama dialect. This small graceful antelope is widespread in Africa south of the Sahara.

Chousingha or Four-horned Antelope   *Tetracerus quadricornis*
tetra (Gr) four   *keras* (Gr) the horn of an animal   *quadri-* (L) prefix meaning four   *cornu* (L) the horn of an animal. *Chousingha* is a Hindustani name meaning 'four horns'. Inhabiting India.

Nilgai or Nilgau   *Boselaphus tragocamelus*
bos (L) an ox   *elaphos* (Gr) a deer   *tragos* (Gr) a he-goat   *kamēlos* (Gr) a camel; a peculiar mixture of names for this rather beautiful antelope with a pale blue coat. Nilgau is from the Hindustani word *nil*, meaning blue, and the Persian word *gaw*, a cow. Inhabiting India and Tibet.

Bushbuck   *Tragelaphus scriptus*
tragos (Gr) a he-goat   *elaphos* (Gr) a deer; in most cases tragos, a goat, is used in the sense of antelope   *scribo* (L) I write, thus *scriptum*, something written; it has conspicuous markings of white stripes and spots on a reddish-brown coat. This is the small member of the genus *Tragelaphus* and it inhabits almost the whole of Africa south of the Sahara.

Sitatunga *T. spekei*
Named after Captain J. H. Speke (1827–1864) the explorer of Central
Africa. He made expeditions with J. A. Grant and also Sir Richard
Burton. The name *sitatunga* is from an archaic Bantu language of
Rhodesia. Inhabiting a large area of Africa including Rhodesia,
Zaire, and ranging to the east and west coasts.

Nyala *T. angasi*
George French Angas (1822–1886) was an English explorer, artist
and zoologist. He went to South Africa in 1846 and made drawings
of men and various animals. *Nyala* is a Swahili name for this bushbuck
inhabiting south-east Africa.

Mountain Nyala *T. buxtoni*
The first specimen was found by a Mr Ivor Buxton in the Abyssinian
highlands in 1909. This nyala is found only in Ethiopia, formerly
Abyssinia.

Lesser Kudu *T. imberbis*
imberbis (L) beardless; referring to the absence of the throat mane.
Inhabiting Somalia and north-eastern Kenya.

Greater Kudu *T. strepsiceros*
strepho (Gr) I twist, so *strepsis*, a twisting   *keras* (Gr) the horn of an
animal; it has remarkable spirally twisted horns   *kudu*, or *koodoo*, is
the Hottentot name for this antelope. It inhabits Ethiopia, eastern
Africa and South Africa.

Common Eland *Taurotragus oryx oryx*
tauros (Gr) a bull   *tragos* (Gr) a he-goat   *orux* (Gr) a gazelle or
antelope. This large antelope lives in the southern part of Africa.

Giant Eland *T. o. derbianus*
-anus (L) suffix meaning belonging to. The Thirteenth Earl of Derby
(1775–1851) was formerly the Hon. E. S. Stanley, President of the
Zoological Society of London in 1831 (not to be confused with Sir
Henry Morton Stanley, famous for exploring Africa and his meeting
with Livingstone at Ujiji in 1871). This eland is even bigger than the
Common Eland (above), and inhabits Sudan, part of western Africa
and the northern part of Zaire.

Bongo *T. euryceros*
eurus (Gr) broad, widespread   *keras* (Gr) the horn of an animal;

reference to the long and elegant horns: bongo is an African native name. It inhabits a belt of forest country from Guinea in the west, across Central Africa and northern Zaire, to Kenya.

South African Oryx or Gemsbok   *Oryx gazella gazella*
orux (Gr) a gazelle or antelope   *gazella* (New L) from *ghazāl* (Ar) a wild goat. Gemsbok is from *gemse* (Ger) the chamois, and *bok* (Du) a buck. Inhabiting the desert areas of South Africa.

East African Oryx or Beisa   *O. g. beisa*
Beisa is from *bezā*, Amharic, an Ethiopian language from the Amhara district where this oryx lives.

Tufted Oryx   *O. g. callotis*
kallos (Gr) beauty; can mean a beautiful object   *ous* (Gr), genitive *ōtos*, the ear; a reference to the long tufts of hair on the ears. Inhabiting the southern part of Somaliland and part of northern Kenya.

Scimitar Oryx   *O. dammah*
damma (L) a fallow deer, an antelope; also *dammar* (Ar) a sheep. Sometimes known as the Scimitar-horned Oryx, the horns are sabre-shaped and bent downwards. It inhabits the southern part of the Sahara.

Arabian or White Oryx   *O. leucoryx*
orux (Gr) a gazelle or antelope   *leukos* (Gr) white. This oryx is near to extinction and recently the Sultan of Muscat and Oman has prohibited their hunting. Some specimens have been taken to Arizona, USA, for protection. It inhabits the desert in the south-west of Saudi Arabia.

Addax   *Addax nasomaculatus*
addax (L) a wild animal with crooked horns, derived from an African word   *nasus* (L) the nose   *macula* (L) a spot, a mark   *-atus* (L) suffix meaning provided with; this refers to brown patches on the nose; it has spirally twisted horns. Becoming rare due to hunting, it is still found in parts of the Sahara.

Waterbuck   *Kobus ellipsiprymnus*
kobus (New L) from *koba*, an African native name   *ellipēs* (Gr) wanting, defective; an ellipse is a shape deviating from a circle, i.e. a defective circle'   *prumnos* (Gr) the hind part; this refers to a conspicuous ellipse-shaped white ring on the rump. Widespread in most of Africa south of the Sahara, but not the extreme south of South Africa.

Defassa Waterbuck   *K. (e.) defassa*
defassa (New L) probably from a native name. A possible subgenus
of *K. ellipsiprymnus*. Widespread in areas not inhabited by the common
waterbuck, as far as the west coast of Africa south of the Sahara.

Buffon's Kob   *K. kob*
Named in honour of the great French naturalist Comte de Buffon
(1707-1788). Inhabiting Africa south of the Sahara and also part of
Angola.

Puku   *K. vardoni*
Named in honour of Major Frank Vardon, an English elephant
hunter, and a friend of Livingstone when in Africa about the year
1850; he wrote the first scientific paper on the Tsetse Fly. Puku is an
African native name. This waterbuck inhabits the Zambezi area.

Lechwe   *K. leche*
Lechwe is a name of Bantu origin meaning an antelope. Widespread
in the Zambezi area, and parts of Angola, Botswana and South West
Africa.

Mrs Gray's or Nile Lechwe   *K. megaceros*
megas (Gr) big, wide   *keras* (Gr) the horn of an animal. Named *Kobus
maria* in 1859 by Dr J. E. Gray FRS (1800-1875), keeper at the British
Museum, in honour of his wife Maria E. Gray (1787-1876); but the
name had to be replaced by the earlier *K. megaceros* (1855). Mrs Gray
was herself a talented artist of molluscs, and the popular English name
has survived. This lechwe is a rare species inhabiting the Bahr-el-
Ghazal and White Nile area in southern Sudan.

Common Reedbuck   *Redunca arundinum*
reduncas (L) bent backwards, curved; most people would describe the
horns as 'bent forward', though they do start at an angle backwards
from the head   *arundo (harundo)* (L) a reed, hence *arundinum*, pertaining to reeds; they are usually found in the vicinity of water. Inhabiting
the southern half of Africa.

Bohor Reedbuck   *R. redunca*
Bohor is an Ethiopian (Amharic) name for this reedbuck. Inhabiting
central parts of Africa from west coast to east coast.

Mountain Reedbuck   *R. fulvorufula*
fulvus (L) tawny   *rufus* (L) red, so *rufulus*, somewhat red, reddish
it is usually more tawny than red. Not always found near water in it

mountain habitat, it lives in scattered areas in Nigeria, Cameroun, Kenya and eastern South Africa.

Impala   *Aepyceros melampus*
aipos (Gr) high, lofty   *keras* (Gr) the horn of an animal; referring to the long lyre-shaped horns of the male   *melas* (Gr) black   *pous* (Gr) the foot   *melampous* (Gr) black-footed; sometimes used in ancient Greece as a proper name, Blackfoot, but in this case an allusion to the tufts of black hair covering a gland on the heel of the hind legs. *Impala* is a Zulu name. It inhabits a large area of southern Africa extending as far north as Uganda.

Roan Antelope   *Hippotragus equinus*
hippos (Gr) a horse   *tragos* (Gr) a he-goat   *equus* (L) a horse and so *equinus*, relating to horses; the body is similar to that of a horse, with a reddish-brown coat. The range is widespread in Africa south of the Sahara.

Sable Antelope   *H. niger niger*
niger (L) black, dark coloured; the coat of both male and female is not jet-black, though in older bulls the coat becomes black; in heraldry sable means black. It inhabits the forests of the southern half of Africa but not found in the extreme south.

Giant Sable Antelope   *H. n. variani*
The type specimen was obtained in 1913 by H. F. Varian, chief engineer on the Benguela Railway, Angola. This large antelope is very rare, and only found in Angola.

Blesbok   *Damaliscus dorcas*
damalis (Gr) a young cow, a heifer   *-iscus* (L) diminutive suffix *dorkas* (Gr) an antelope or gazelle; a small antelope with a white patch on the forehead and nose. The name blesbok comes from *bles* (Du) a mark or blaze, and *bok* (Du) a buck, alluding to the white blaze on the forehead. A rare antelope found only in South Africa, and protected in some game reserves.

Hunter's Hartebeest   *D. hunteri*
H. C. V. Hunter (1861–1934) was a big game hunter and zoologist; he discovered this antelope in 1888 about 240 km (150 miles) up the Tana River in Kenya.

A rare hartebeest found only in a small area in southern Somalia and northern Kenya. The name is a combination of 'hart' and 'beast'; the hart is a male deer and the hind is a female.

Bontebok   *D. pygargus*
pugē (Gr) the rump, buttocks   *argos* (Gr) shining, bright   *pugargos*
(Gr) white-rump; it has a conspicuous white rump and tail   *bont* (Du)
particoloured   *bok* (Du) a buck. Found only in the south-west of
South Africa, it is now extremely rare.

Korrigum or Topi   *D. korrigum*
Korrigum is from *kargum* (Kinuri) the language of a tribe living near
Lake Chad, Niger   *topi* is a native name of Mande origin, akin to
ndope, an antelope. Another rare antelope from West Africa.

Common Hartebeest   *Alcelaphus buselaphus*
alkē (Gr) the elk   *elaphos* (Gr) a deer   *bous* (Gr) a bullock or cow.
For hartebeest, see above. Inhabiting a wide area in central Africa
stretching from the west coast to the east.

Lichtenstein's Hartebeest   *A. lichtensteini*
M. H. C. Lichtenstein (1780-1857) was Director of Zoology at the
Berlin Museum in 1815; he made an expedition to South Africa in
1804. This hartebeest lives in Tanzania, Mozambique and Zambia.

White-tailed Gnu or Wildebeest   *Connochaetes gnou*
konnos (Gr) the beard   *khaitē* (Gr) flowing hair, can mean a mane; a
reference to the conspicuous beard and mane   *gnou* is a Hottentot
name for these peculiar and rather ugly antelopes. This one is a very
rare species, and probably now only existing in protected areas in
Namibia and in zoos.

Southern Brindled or Blue Gnu   *C. taurinus taurinus*
taurus (L) a bull; taurinus, like a bull. Inhabiting the southern part
of Africa.

White-bearded Gnu   *C. t. albojubatus*
albus (L) white   *juba* (= *iuba*) (L) the mane of an animal, hence
jubatus, having a mane; the long hairs on the throat and beneath the
neck of an animal are sometimes known as the mane. Inhabiting East
Africa and parts of South Africa.

Indian Antelope or Blackbuck   *Antilope cervicapra*
antholops (Gr) an antelope: (see page 212)   *cervus* (L) a deer   *capr*
(L) a she-goat. The only species in the genus *Antilope*, it is widesprea
in India.

Springbok *Antidorcas marsupialis*

anti- (Gr) against, opposed to; in composition can mean resemblance to the word that follows *dorkas* (Gr) a gazelle; a reference to their likeness to *Gazella*; *marsupium* (L) a pouch, a pocket *-alis* (L) suffix meaning similar to, like; on the back there is a long pocket of skin which can be opened at will, revealing a conspicuous mass of white hair; this serves as a danger signal. Inhabiting semi-desert areas of South Africa.

Goa or Tibetan Gazelle *Procapra picticaudata*

pro (L) before *capra* (L) a she-goat; suggesting the ancestral or original type of *Capra*; *pictus* (L) painted *cauda* (L) the tail of an animal *-atus* (L) suffix meaning provided with; 'having a painted tail'; a reference to its conspicuous white tail and rump; goa is from *dgoba* (Tibetan) a gazelle. It inhabits Tibet and neighbouring areas of Asia.

Mongolian Gazelle *P. gutturosa*

guttur (L) the throat *-osus* (L) suffix meaning full of; the males have an enlargement of the neck and throat during the mating season which resembles a goitre. Inhabiting Mongolia and neighbouring areas of Asia.

Grant's Gazelle *Gazella granti*

ghazal (Ar) a wild goat *-ellus* (L) diminutive suffix; named after Colonel J. A. Grant (1827–1892) who explored central Africa during the years 1860 to 1862, and on one occasion with J. H. Speke. This gazelle inhabits Kenya and Tanzania.

Mountain Gazelle *G. gazella*

Inhabiting the mountains of North Africa and to some extent the northern part of the Sahara; also Arabia and south-western Asia.

Dorcas Gazelle *G. dorcas*

dorkas (Gr) a gazelle. Inhabiting northern Africa including parts of the Sahara, and Saudi Arabia.

Thomson's Gazelle *G. thomsoni*

Named after Joseph Thomson (1858–1895) who explored the area of the Central African Lakes during the years 1879 to 1883. These beautiful little gazelles live in that area, particularly in the Serengeti National Park. They are always known locally by the popular name 'Tommies'.

Heuglin's Gazelle   *G. tilonura*
ptilon (Gr) a feather, down; also *tilos* (Gr) anything pulled or shredded, flock, down   *oura* (Gr) the tail; 'feathery-tailed'. M. T. von Heuglin (1824–1876) was a German zoologist who was in Africa from 1851 to 1864. This gazelle is becoming rare; it is still found in a small area in the eastern part of Sudan and northern Ethiopia.

Cuvier's Gazelle or Edmi   *G. cuvieri*
Baron G. L. Cuvier (1769–1832) was the famous French comparative anatomist and Professor of Natural History   *edmi* or *idmi* is an Arabic local native name. This gazelle is now very rare; it inhabits the northern part of the Sahara.

Pelzeln's Gazelle   *G. pelzelni*
A. von Pelzeln (1825-1891) was a zoologist and Custodian of the Vienna Museum from 1859 to 1883. This gazelle inhabits Somaliland and the eastern part of Ethiopia.

Goitred or Persian Gazelle   *G. subgutturosa*
sub (L) below   *guttur* (L) the throat   *-osus* (L) suffix meaning full of the male has an enlargement of the neck and throat during the mating season which resembles a goitre. In fact, a goitre is a swelling of the throat caused by enlargement of the thyroid gland and has no connection with the enlargement of the throat of the gazelle. This gazelle inhabits a large area in central Asia including part of Iran at the western end of the range.

Clarke's Gazelle or Dibatag   *Ammodorcas clarkei*
ammos (Gr) sand, a sandy place   *dorkas* (Gr) a gazelle; a reference to its dry, sandy habitat; a specimen was obtained by T. W. H. Clark in 1890 in the Marehan country, in the southern part of Somalia Dibatag is from the Somali name *dabatag*. It inhabits Somaliland and Ethiopia.

Waller's Gazelle or Gerenuk   *Litocranius walleri*
lithos (Gr) stone   *kranion* (Gr) the upper part of the skull; 'stone skull'; a reference to the skull which is almost solid bone at the base of the horns; named after the Rev. H. Waller (1833–1901) who was a missionary in Africa and a friend of the famous explorer Dr Livingstone. Gerenuk is from the Somali *garanug*, a long-necked gazelle. It inhabits Somaliland and part of neighbouring Ethiopia and Kenya

Saiga Antelope   *Saiga tatarica*

Saiga is a Russian name for an antelope   Tatary is an area in the
eastern part of European Russia   *-icus* (L) suffix meaning belonging
to. It has been in danger of extinction but can still be found in Kazakhstan and neighbouring areas. It has a dome-shaped nose, even larger
than that of *Pantholops* (below).

Tibetan Antelope or Chiru   *Pantholops hodgsoni*

pas (neuter *pan*) (Gr) all   *antholops* (Gr) an antelope; a strange name.
T. S. Palmer, in his standard work *Index Generum Mammalium*, quotes
'The vulgar old name for the unicorn' (Hodgson). He goes on to
explain that when seen in profile the two horns appear like one, which
has given rise to the belief that the animal is the unicorn antelope
mentioned by the Abbé Huc. The Mr B. H. Hodgson FRS (1800–
1894) was a biologist who lived in Nepal during the years 1833 to
1843. Chiru is probably a local native name in Tibet. This goat-like
antelope has a peculiar dome-shaped nose, the purpose of which is still
under discussion among zoologists. It inhabits the high mountains of
Nepal and Tibet.

Grey Goral   *Nemorhaedus goral goral*

nemus (L), genitive *nemoris*, a grove, a forest   *haedus* (L) a young goat,
a kid; an allusion to its habitat in mountainous and woody regions
goral is a native name from eastern India. This goat-antelope covers
a vast range including Afghanistan, Tibet, eastern Siberia and parts
of China and Korea.

Red Goral   *N. g. cranbrooki*

The Fourth Earl of Cranbrook (born 1900) is a zoologist who was in
Burma in 1930, and was a Trustee of the British Museum of Natural
History in 1963. This goral has a shaggy coat of bright fox-red; it lives
in the mountains of northern Burma and Assam.

Maned Serow   *Capricornis sumatraensis*

capra (L) a she-goat   *cornu* (L), genitive *cornus*, the horn of an animal;
a reference to the goat-like horns   *-ensis* (L) suffix meaning belonging.
It is not confined to Sumatra, and ranges over a wide area to the north
including Pakistan, Tibet and most of China. Serow is a name used
by the Lapchas, who inhabit Sikkim in the Himalayas, and now used
for the other goat-like animals in this genus.

Japanese Serow   *C. crispus crispus*
crispus (L) curly-headed. This serow is confined to Japan and neighbouring islands.

Formosan Serow   *C. c. swinhoii*
R. Swinhoe FRS (1836–1877) was at one time in the British Consular Service in China. This serow lives on the island of Formosa (Taiwan).

Rocky Mountain Goat   *Oreamnos americanus*
oros (Gr), genitive *oreos*, a mountain   *amnos* (Gr) a lamb. In the Rocky Mountains of Canada and the northern part of the USA, this aberrant goat-antelope, a relative of the Chamois, can still be found though it is becoming rare; some are protected in national parks.

Chamois   *Rupicapra rupicapra*
rupes (L), genitive *rupis*, a rock or cliff   *capra* (L) a she-goat. *Chamois* is French for a wild goat. This agile rock-climber lives high up in the mountains of Europe and western Asia.

Wild Goat   *Capra aegagrus*
capra (L) a she-goat   *aix* (Gr), genitive *aigos*, a goat   *agrios* (Gr) living in the fields; of animals, wild, hence *aigagros* (Gr) a wild goat. This goat is considered to be the ancestor of the domestic goat. It is sometimes known as the Bezoar Goat, bezoar being a stony substance found in the stomach of some ruminants such as goats and which used to be considered an antidote for all poisons. The word derives from *pādzahr* (Persian), from *pād*, protecting, and *zahr*, poison. Formerly widespread in Asia Minor and the Greek Islands, it is now found only in parts of Turkey, Georgia and Iran.

Alpine Ibex   *C. ibex ibex*
ibex (L) a kind of goat, a chamois. Inhabiting the mountains of southern Europe.

Siberian Ibex   *C. i. sibirica*
-icus (L) suffix meaning belonging to. Inhabiting parts of Siberia Tibet and China.

Nubian Ibex   *C. i. nubiana*
-anus (L) suffix meaning belonging to; Nubia is a tract of country in northern Africa with no precise limits, lying between Egypt and Sudan. This ibex also ranges through Saudi Arabia and north Syria.

Severtzow's Ibex *C. i. severtzovi*
Professor N. A. Severtzow (1827–1885) was a scientist who explored Central Asia. This ibex lives in the western part of the Caucasus Mountains.

Walia or Abyssinian Ibex *C. i. walie*
Walia was originally an Ethiopian native name. This ibex is in danger of extinction and attempts are being made to preserve it. A fine animal, it is confined to a small area in the mountains of the eastern part of Ethiopia.

Caucasian Ibex or Tur *C. caucasica*
-*icus* (L) suffix meaning belonging to *tur* (Russ) a Caucasian goat. Inhabiting the Caucasus Mountains in southern Russia, along the border with Georgia.

Pyrenees Ibex *C. pyrenaica pyrenaica*
This ibex is now extinct but is included because it is the nominate subspecies.

Spanish Ibex *C. p. hispanica*
icus (L) suffix meaning belonging to. A rare ibex in danger of extinction, it inhabits the Sierra Nevada mountains in Southern Spain.

Queen Victoria's Ibex *C. p. victoriae*
Named in honour of Queen Victoria Eugenie (formerly Princess Ena of Battenberg) and the god-daughter of Queen Victoria; she married King Alfonso XIII of Spain in 1906. Another very rare ibex in danger of extinction, inhabiting Spain.

Markhor *C. falconeri*
Hugh Falconer (1808–1865) was a Scottish palaeontologist and botanist in India. Markhor is a name that derives from the Persian *mār*, a snake, and *khor*, eating. This is a strange name as goats are vegetarians and no actual case of the markhor eating snakes has been recorded, though they have been known to kill snakes. It inhabits the mountains of Central Asia, originally including Afghanistan and Persia, though it is now probably extinct in those countries.

There are varying opinions as to the number of species in *Capra* and *Ovis*. The German systematist Th. Haltenorth reduces *Ovis* to one, namely *ammon*, whilst we follow Desmond Morris and others with seven. The Russian scientist Nabonov recognised as many as nine in the Old World alone.

Barbary Sheep or Aoudad *Ammotragus lervia*
ammos (Gr) sand   *tragos* (Gr) a goat; this is a reference to the colour
of its coat   *lervia*, from the wild sheep of northern Africa called
'Fishtall' or 'Lerwee' by the Rev T. Shaw in his *Travels and Observa-
tions* relating to several parts of Barbary and the Levant. It was named
by Pallas in 1777. *Aoudad* is a name used by the Berbers, a people of
northern Africa. It inhabits various parts around the Sahara, particularly Kordofan in Sudan, and Barbary, which is the old name for the
belt of land north of the Sahara stretching from Egypt to the Atlantic.
Berber is from the Arabic *Barbar*.

Himalayan Tahr *Hemitragus jemlahicus*
hēmi (Gr) half   *tragos* (Gr) a goat; meaning 'something like a goat'
a reference to the absence of a beard and the animal having some of
the habits and characters of a goat   *jemlah*, probably from *him*
(Sanskrit) snow, and *alaya*, an abode; hence also Himalaya   *-icus* (L)
suffix meaning belonging to. Tahr is from *thār*, the Nepalese name for
this wild goat inhabiting the Himalayas.

Arabian Tahr *H. jayakari*
Named after Surgeon Colonel A. S. G. Jayakar; he collected in the
Persian Gulf, chiefly birds, from 1878. During the years 1885 to 1890
he presented collections to the British Museum. Inhabiting the Muscat
and Oman area of south-eastern Arabia.

Nilgiri Tahr *H. hylocrius*
hulē (Gr) a wood, a forest   *krios* (Gr) a ram. The Nilgiri Hills are in
the south-eastern part of India.

Blue Sheep or Bharal *Pseudois nayaur*
pseudēs (Gr) false   *ois* (Gr) a sheep; referring to the absence of facial
glands, and the character of the tail, which makes this genus resemble
the goats more than the sheep   *nayaur* is a native name for this wild
sheep, probably from the Nepali word *nahūr*. *Bharal* is a Hindi name.
The coat is grey, becoming more blue in winter in the case of juveniles.
It inhabits the Himalayas from India to China.

Sheep *Ovis aries*
ovis (L) a sheep   *aries* (L) a ram. This is the domestic sheep, and there
are now over 400 different breeds. It was first domesticated many
thousands of years ago and the actual ancestry is uncertain.

Mouflon *O. musimon*
musimo (L) an animal of Sardinia   *mouflon* (Fr) a Sardinian wild

sheep. Considered to be one of the ancestors of the domestic sheep, it is only found on the islands of Corsica and Sardinia.

Laristan Sheep   *O. laristanica*
Lar, formerly Laristan, is a town and district in Iran, bounded on the south by the Persian Gulf. This sheep is confined to the southern part of Iran.

Urial or Red Sheep   *O. orientalis*
-*alis* (L) suffix meaning relating to. Urial is from *hureāl* (Punjabi) a Himalayan wild sheep. The coat is a brown to red colour. Inhabiting the Himalayas in Afghanistan and ranging south to Baluchistan.

Argali   *O. ammon*
Ammon or Amen was an Egyptian deity, usually represented in human form with a ram's head. *Argali* is a Mongolian name for this sheep; it covers a wide area throughout eastern Central Asia, including parts of Mongolia, Tibet and the Gobi Desert.

Rocky Mountain Sheep or Bighorn   *O. canadensis*
-*ensis* (L) suffix meaning belonging to; it is not confined to Canada and ranges through mountainous areas as far south as Mexico; it has massive horns.

Dall's or White Sheep   *O. dalli*
William H. Dall (1845–1927) was an American zoologist; this sheep was discovered in 1884 and named in his honour. The coat is grey to white and it inhabits Alaska and the western part of Canada.

Takin   *Budorcas taxicolor*
bu (= *boo, bous*) (Gr) an ox   *dorkas* (Gr) a gazelle; a gazelle-like ox *taxus* (New L) a badger; badger-coloured, a yellowish grey. *Takin* is a Tibeto-Burman name for this animal which is related to the musk-oxen; it inhabits Tibet and neighbouring areas.

Musk Ox or Musk Sheep   *Ovibos moschatus*
ovis (L) a sheep   *bos* (L) an ox; it has features in common with the ox and the sheep   *moschatus* (New L) musky; it has preorbital glands that secrete a musky odour. The heavy long-haired coat probably constitutes the best 'warmth preserver' among land-living animals; very necessary for this ox for protection against the intense cold of Greenland, northern Canada and Alaska.

Tamarau   *Anoa mindorensis*
Mindoro is an island in the Philippines   -*ensis* (L) suffix meaning

belonging to *anoa* is the Celebes native name for this buffalo. Tamarau (Tagalog) is from the language of a people of Luzon, in the Philippines.

Anoa   *A. depressicornis*
de (L) down from   *presso* (L) I press   *cornu* (L) a horn; the horns are short and depressed backwards. Inhabiting Celebes, Indonesia.

Water Buffalo   *Bubalus arnee*
boubalos (Gr) a buffalo   *arnee* is from the Hindi native name *arnā*. Although domesticated there are still some wild herds in Borneo, Malaya, Thailand and India.

African Buffalo   *Syncerus caffer caffer*
sun (Gr) together   *keras* (Gr) the horn of animals; an allusion to the horns which are close together at the base   *cafer* (New L) of Caffraria (or Kaffraria), the country of the Kaffirs. It is widespread in the southern half of Africa but not found now in the extreme south.

Forest or Dwarf Buffalo   *S. c. nanus*
nanus (L) a dwarf; this small buffalo lives in forested areas of western Africa including Congo, Cameroun and Nigeria and ranging west to Guinea.

Banteng   *Bos javanicus*
bos (L) an ox   *-icus* (L) suffix meaning belonging to; it inhabits Java, Borneo and part of southern Asia. *Banteng* is a Malayan name.

Gaur   *B. gaurus*
Gaur is the Hindustani name for this large ox inhabiting India, Burma, and other parts of southern Asia.

Kouprey   *B. sauveli*
Only recently discovered and first described by Professor A. Urbain, the French zoologist, in 1937. He had previously seen, for the first time, the horns of this ox in the home of Dr Sauvel, a veterinary surgeon in Cambodia. Thus, he named it in his honour. Kouprey is the native name in Cambodia. A rare animal, restricted to an area in the Mekong River valley in Cambodia, and part of Laos and Vietnam.

Aurochs   *B. primigenius*
primigenus (L) original, primitive; for many years there has been discussion about the probable ancestors of domestic cattle. Zoological

research in recent years shows that they are probably descended from this ox. The name aurochs is derived from Old High German. It is now extinct, but was known in Europe up to the sixteenth century.

Zebu (Domestic) *B. indicus*
-icus (L) suffix meaning belonging to; 'of India'; originally from India it exists now only in the domestic form and is widespread in the east. Zebu is a French name and first adopted at a French fair in 1752.

Western Cattle (Domestic) *B. taurus*
taurus (L) a bull. Almost world-wide in farms, ranches, etc.

Wild Yak *B. mutus*
mutus (L) dumb, unable to speak; they cannot 'moo' like normal cattle, but only grunt. Inhabiting high altitudes in Tibet.

American Bison *Bison bison bison*
bison (L) a bison   *bisōn* (Gr) a species of wild ox, the humpbacked ox, bison. In the eighteenth and early nineteenth centuries, there were many millions of these fine animals in North America, but gradually as the white settlers invaded the country and began to build railways the beasts were systematically destroyed for food. Now there are probably none in existence, except in wildlife parks and game reserves. (See Tautonyms, page 13.)

Wood Bison *B. b. athabascae*
The type was named from a region known as Athabaska, in North West Canada, south of the Great Slave Lake. There is a Lake Athabaska and a River Athabaska and a large area is now established as Wood Buffalo National Park, where over ten thousand animals are preserved. This subspecies is larger and darker, and has longer thinner horns.

European Bison *B. bonasus*
bonasus (L) a kind of buffalo. It is sometimes known as the Wisent, from the German word *wisunt*, a bison. Like the American Bison, it is now unknown in the wild state, but many are protected in wildlife parks in Poland and other countries in Europe, and in the Caucasus Mountains.

Appendix

Transliteration of Greek Alphabet

Greek		Name	Modern System	Latin System
A	α	alpha	a	a
B	β	bēta	b	b
Γ	γ	gamma	g	g
Δ	δ	delta	d	d
E	ε	epsīlon	e	e
Z	ζ	zēta	z	z
H	η	ēta	ē	e
Θ	θ	thēta	th	th
I	ι	iōta	i	i
K	κ	kappa	k	c
Λ	λ	lambda	l	l
M	μ	mū	m	m
N	ν	nū	n	n
Ξ	ξ	xī	x	x
O	o	omīcron	o	o
Π	π	pī	p	p
P	ρ	rhō	r	r
Σ	σ ς	sigma	s	s
T	τ	tau	t	t
Y	υ	upsīlon	u	y
Φ	φ	phī	ph	ph
X	χ	chī	kh	ch
Ψ	ψ	psī	ps	ps
Ω	ω	ōmega	ō	o

Bibliography

Buschbaum, R. and Milne, L. J. *Living Invertebrates of the World*, Hamish Hamilton, London.

Clark, R. B. and Panchen, A. L. *Synopsis of Animal Classification*, Chapman and Hall, London.

Jaeger, E. C. *A Source Book of Biological Names and Terms*, Charles C. Thomas, Illinois.

Jeffrey, C. *Biological Nomenclature*, Edward Arnold, London.

Lerwill, C. J. *An Introduction to the Classification of Animals*, Constable, London.

Lord Rothschild, *A Classification of Living Animals*, Longman, London.

Mayr, E. *Principles of Systematic Zoology*, McGraw-Hill.

Morris, D. *The Mammals*, Hodder and Stoughton, London.

Sanderson, I. T. *Living Mammals of the World*, Hamish Hamilton, London.

Savory, T. H. *Latin and Greek for Biologists*, Merrow.

General Index

This index includes only technical terms, names of zoologists, and all animals other than mammals mentioned in the main text (Part 2) of the book; some animals are given with their Latin names in *italics*. Phyla, Subphyla and Classes are indicated THUS.

Index of English Names

This index contains the English names of all mammals in the book. In general, the page number of the main reference is given, but where there is an additional reference of a general or less important nature, then this page number is indicated in *italic* type.

Index of Latin Names

This index contains the Latin names of all mammals. The Phylum, Subphylum, Class, Subclasses, Orders, Suborders and Families are indicated THUS. In general, the page number of the main reference is given, but where there is an additional reference of a general or less important nature, then this page number is indicated in *italic* type.

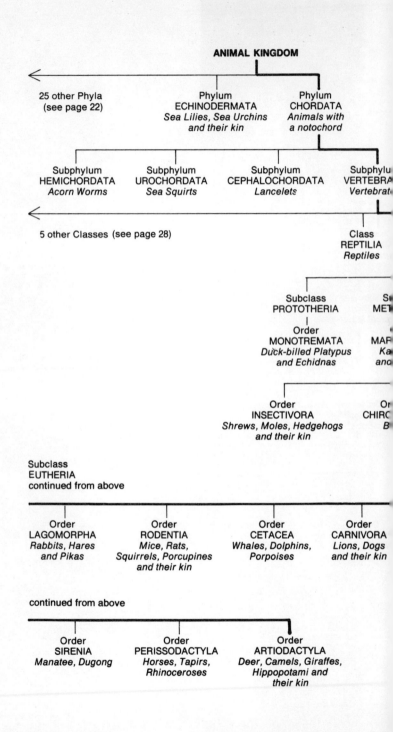

ANIMAL KINGDOM

25 other Phyla
(see page 22)

Phylum
ECHINODERMATA
*Sea Lilies, Sea Urchins
and their kin*

Phylum
CHORDATA
*Animals with
a notochord*

Subphylum
HEMICHORDATA
Acorn Worms

Subphylum
UROCHORDATA
Sea Squirts

Subphylum
CEPHALOCHORDATA
Lancelets

Subphylu
VERTEBRA
Vertebrat

5 other Classes (see page 28)

Class
REPTILIA
Reptiles

Subclass
PROTOTHERIA

S
MET

Order
MONOTREMATA
*Duck-billed Platypus
and Echidnas*

MAR
*Ka
and*

Order
INSECTIVORA
*Shrews, Moles, Hedgehogs
and their kin*

Or
CHIRO
B

Subclass
EUTHERIA
continued from above

Order
LAGOMORPHA
*Rabbits, Hares
and Pikas*

Order
RODENTIA
*Mice, Rats,
Squirrels, Porcupines
and their kin*

Order
CETACEA
*Whales, Dolphins,
Porpoises*

Order
CARNIVORA
*Lions, Dogs
and their kin*

continued from above

Order
SIRENIA
Manatee, Dugong

Order
PERISSODACTYLA
*Horses, Tapirs,
Rhinoceroses*

Order
ARTIODACTYLA
*Deer, Camels, Giraffes,
Hippopotami and
their kin*